Improving International Irrigation Management with Farmer Participation

Studies in Water Policy and Management

Charles W. Howe, General Editor

Improving International Irrigation Management with Farmer Participation: Getting the Process Right, Norman Uphoff

Municipal Water Demand: Statistical and Management Issues, Clive Vaughan Jones, John J. Boland, James E. Crews, C. Frederick DeKay, and John R. Morris

Flood Plain Land Use Management: A National Assessment, Raymond J. Burby and Steven P. French with Beverly A. Cigler, Edward J. Kaiser, David H. Moreau, and Bruce Stiftel

Natural Radioactivity in Water Supplies, Jack K. Horner

Energy and Water Management in Western Irrigated Agriculture, edited by Norman K. Whittlesey

Irrigation Management in Developing Countries: Current Issues and Approaches, edited by K. C. Nobe and R. K. Sampath

Irrigation Investment, Technology, and Management Strategies for Development, edited by K. William Easter

Watershed Resources Management: An Integrated Framework with Studies from Asia and the Pacific, edited by K. William Easter, John A. Dixon, and Maynard M. Hufschmidt

About the Book and Author

Traditionally, indigenous irrigation in many countries of Asia, Africa, and Latin America has been managed quite well by water users, who design, build, operate, and maintain often sophisticated, but usually small-scale, systems. More recently, in connection with large-scale development programs and government-managed schemes, the planned introduction of water user associations is increasing with some notable success. Dr. Uphoff brings together lessons from indigenous and government-managed irrigation systems worldwide and proposes workable approaches to participatory irrigation management in both the large- and small-scale contexts.

Norman Uphoff is professor of government and chairman of the Rural Development Committee in the Center for International Studies at Cornell University.

Improving International Irrigation Management with Farmer Participation

Getting the Process Right

Norman Uphoff

Studies in Water Policy and Management, No. 11

Westview Press / Boulder and London

Studies in Water Policy and Management

This Westview softcover edition is printed on acid-free paper and bound in softcovers that carry the highest rating of the National Association of State Textbook Administrators, in consultation with the Association of American Publishers and the Book Manufacturers' Institute.

Published in 1986 in the United States of America by Westview Press, Inc.; Frederick A. Praeger, Publisher; 5500 Central Avenue, Boulder, Colorado 80301

Library of Congress Cataloging-in-Publication Data
Uphoff, Norman Thomas.
Improving international irrigation management with farmer participation.
Bibliography: p.
1. Irrigation—Management—Citizen participation.
I. Title.
HD1714.U64 1986 333.91'3 86-23414
ISBN 0-8133-7330-1

Composition for this book was provided by the author.
This book was produced without formal editing by the publisher.

Printed and bound in the United States of America

The paper used in this publication meets the requirements of the American National Standard for Permanence of Paper for Printed Library Materials Z39.48-1984.

6 5 4 3 2 1

Contents

LIST OF TABLES AND FIGURES	xi
ACKNOWLEDGMENTS	xiii
I. INTRODUCTION	
1. IRRIGATION AS A SOCIO-TECHNICAL PROCESS	3
System Management by Agencies and by Users	7
Farmer Organization and Participation	8
Empirical Bases for Improving Farmer Participation	11
2. WHY FARMER PARTICIPATION? CONTRIBUTIONS TO IRRIGATION MANAGEMENT	19
Objectives in Irrigation Management	19
Benefits of Farmer Participation	24
Costs of Farmer Participation	26
Comparisons of Benefits and Costs	27
II. ANALYZING FARMER ORGANIZATION AND PARTICIPATION	
3. WHAT KINDS OF PARTICIPATION? ACTIVITIES IN IRRIGATION MANAGEMENT	37
Water Use Activities	42
Control Structure Activities	45
Organizational Activities	46
External Organizational Activities	53

4. WHERE CAN PARTICIPATION OCCUR? LEVELS OF OPERATION AND ORGANIZATION IN IRRIGATION SYSTEMS 59

Types and Levels of Irrigation Systems 60
Rationale for Analyzing Levels of Operation and Organization 67
Relation Between Number of Levels and System Size 69
Irrigation Groups 72
Higher Levels of Organization 74

5. WHO PARTICIPATES? USER ROLES IN IRRIGATION MANAGEMENT 79

Specialization in Participation 81
Incentives for Participation 83
Mobilizing Leadership for Participation 86
User Groups vs. Local Government 87
Participation of Women 89

6. THE CONTEXT OF PARTICIPATION: FACTORS IN THE ENVIRONMENT 93

Historical Factors 93
Physical and Economic Factors 97
Socio-Cultural and Political Factors 102

III. SUPPORTING FARMER ORGANIZATION AND PARTICIPATION

7. POLICY CONSIDERATIONS 109

Support 111
Experimentation, Phasing and Flexibility 112
Building From Below 114
Existing Organizations 115
Ownership 117
Non-Political Associations 118
Bureaucratic Reorientation 120
Compatibility of Objectives 121

8. CHOICES IN ORGANIZATIONAL DESIGN 127

Scope of Activity 128
Size and Structure 130
Membership and Decision-Making 131
Leadership and Responsibilities 133
Legal Basis 135
Development Strategy 137
Channels for Implementation 138

9. IMPROVING AGENCY RELATIONS WITH FARMERS 145

Agency Incentives 146
Technical Cooperation 148
Special Roles 149
Agency Style 152
Agency Organization 152
Concluding Observations 153

ANNEXES

1. Criteria of Irrigation Management Performance 161

2. Analysis of Irrigation Activities 165

3. Irrigation Management Roles 168

REFERENCES 177

INDEXES

Author Index 201

Country/Case Index 205

Subject Index 209

Tables and Figures

TABLES

1. Cases Analyzed 14

2. Relation Between Structure-Focused and Water-Focused Activities 39

FIGURES

1. Matrix of Irrigation Management Activities 42

2. Schematic Comparison of Irrigation Structures by Number of Levels of Operation and Possible Organization 66

3. Water User Knowledge Base for Irrigation Management 94

Acknowledgments

The review of the literature and analysis offered here were undertaken to provide the empirical and conceptual bases for a paper on Improving Policies and Programs for Farmer Organization and Participation in Irrigation Water Management. Preparation of that paper as well as this work was supported by the Water Management Synthesis II Project, funded by USAID and administered by the Consortium for International Development. The policy paper, published at the end of 1985, was written with Ruth Meinzen-Dick (Rural Sociology) and Nancy St. Julien (Regional Planning) who assisted with this state-of-the-art study during much of the last two years.

This undertaking was enriched by discussions with colleagues in the Cornell Irrigation Studies Group: Benjamin Bagadion, Jr., Bryan Bruns (who also did some of the initial case study analyses), E. Walter Coward, Jr., John Duewel, Barbara Lynch, and Ujjwal Pradhan (Rural Sociology); Edward Martin and C. M. Wijayaratna (Agricultural Economics); Gerard Finin and Ruth Yabes (Regional Planning); Chris Wensley and Robert Yoder (Agricultural Engineering); and Milan Rodrigo (Communication Arts).

Helpful comments on the manuscript were given by Bagadion, Coward, Finin, Lynch, Meinzen-Dick and St. Julien as well as by Jayantha Perera, Deputy Director of the Agrarian Research and Training Institute in Sri Lanka who was visiting Cornell during the draft's completion. Dr. Douglas J. Merrey, at the time Senior Social Science Advisor in the Office of Rural and Institutional Development, Bureau for Science and Technology, USAID (now with the International Irrigation Management Institute), oversaw this activity for the Water Management Synthesis Project and made many helpful inputs.

Debi Ostrander and Virginia Hicks in the Center for International Studies at Cornell provided valuable logistical support, and Jane Kellogg did the invaluable job of word processing the manuscript through its several stages. All of their assistance and advice are gratefully acknowledged.

This effort benefited from the deliberations of an international seminar on "Community Responses to Irrigation" organized by P. K. Bardhan and myself for the South Asia Committee of the Social Science Research Council, New York, and held in cooperation with the Indian Institute of Management in Bangalore, January 3-6, 1984. This seminar provided fruitful opportunity to discuss and compare case experiences in some depth.

Participants included Dr. Merrey as well as two members of the Ford Foundation staff in New Delhi who have contributed greatly to the conceptualization of irrigation systems management, Robert Chambers and Gilbert Levine. I would like to express my appreciation to them and to the other seminar participants, whose papers are included in the Bibliography. The report of the seminar is found in WAMANA, the water management newsletter published by the Indian Institute of Management, Bangalore, October, 1984.

In May 1984, a workshop was held at Cornell on "Farmer Participation and Organization for Improved Water Management" under Water Management Synthesis Project auspices. This gave an opportunity to exchange ideas and experience with WMSP colleagues from Colorado State University and Utah State University, as well as other invited workshop participants. The discussion and analysis in Part III of this paper particularly benefited from the deliberations of that workshop.

Norman Uphoff

Recent years have witnessed a marked increase in concern at both national and international levels about the poor performance of irrigation schemes in many developing countries: rapid population growth is intensifying mankind's demands on an increasingly scarce resource which is essential to its livelihood... yet efficiency of water use on many existing systems continues to be very low.

...there has been a growing realization that much of the poor performance [in irrigation systems] stems from fundamental weaknesses in the human processes of planning and management, which no amount of investment in technological hardware is going to overcome on its own.

Anthony Bottrall (1985:8),
emphasis added

Part One

Introduction

1
Irrigation as a Socio-Technical Process

When irrigation structures capture, convey and distribute water to support plant growth, they are quite visible and impressive. Their evident success in sustaining agriculture under conditions where inadequate rainfall would otherwise limit or prevent production makes irrigation appear as a preeminently physical process. However, before there are any physical structures and all the while they are operating, there must be the associated human process to which Bottrall was referring in the quotation introducing this volume. Many management activities are required to make the physical structures perform their intended functions, making irrigation a socio-technical process which combines human and material elements to achieve a more predictable and productive agriculture.

Irrigation systems of course require correct technical planning and design. Dams need to have the material strength and spill facilities to cope with powerful hydraulic forces. Channels must be built with the right capacity and elevations. Control structures should facilitate the timely delivery of appropriate amounts of water. Certainly there is need to "get the engineering right."

Other physical relationships need also to be properly identified and accommodated in plans and operations. Are the soils to be served by the channel system suitable for irrigation? Does the natural drainage have to be supplemented by special facilities? Agronomists and other agricultural scientists have important contributions to make in irrigation management, as do economists, who remind us continually of the need to "get the prices right" if production accomplishments are to match expectations.

But all of these considerations -- engineering, agronomic, economic -- depend for their realization of objectives on the decisions and activities of farmers. The benefits from irrigation are few unless the ultimate water users employ their own labor and capital in ways that make good use of available and anticipated land and water resources. It is users who decide in the final analysis whether the prices are "right" and who judge the suitability of soils and physical structures for growing irrigated crops.

Farmers' considerations of benefit and cost, of what is feasible and desirable, must be taken into account because they control crucial aspects of irrigation system development and management. However, irrigation is a socio-technical matter not just because people are involved in the process. Irrigated agriculture requires a degree of cooperation among water users and with any persons managing the irrigation system at higher levels which sets it apart from rainfed production. Inter-dependence among farmers and between farmers and managers makes effective irrigation a social process in very many respects.

This is not to suggest that irrigation be viewed as more social than technical. Both human and physical aspects interact continually and profoundly in irrigation enterprises, so a hyphenated construct of irrigation as a socio-technical process seems appropriate.[1] What is suggested here is that most approaches to planning and managing irrigation have not adequately conceived and provided for the role of farmers. "Getting the process right" requires irrigation departments and donor agencies to be less preoccupied with technical and economic factors in their planning, policy and operations. The social dimensions of irrigation management have been too often neglected, handled badly, or assumed not to require any special knowledge or expertise. While policy-makers, administrators and technicians are willing to invest heavily in the physical aspects of irrigation -- in designs, surveys, research, experimentation, well-trained staff, etc. -- they commonly make decisions about human organization and behavior that have little empirical basis and that lead to poor returns on investment.

If social scientists ventured into the realms of engineering, agronomy or economics as freely and casually as practitioners of these disciplines engage, often unwittingly, in "social engineering," there would surely be objections. Time and again one sees assumptions being made in plans and policies without much evidence or analysis about how irrigation system managers and water users will act and relate to one another. If there is any thinking about them, it is wishful. Certain behaviors are called for in policies and plans that will match and support decisions already made separately on purely technical or economic grounds.

Perhaps one could justify treating the behavioral aspects of irrigation practically as residual elements that must accomodate planners' ideals if the performance of irrigation systems and sectors in developing countries were more satisfactory. But as

Bottrall's observation introducing this volume notes, there are no grounds for such satisfaction.

The behavioral aspects of irrigation are not limited to water users but include also the activities and attitudes of any agency personnel involved in water management. Irrigation as a socio-technical process encompasses more than just farmer involvement. Where administrators, engineers and technicians play a role in irrigation management, they affect system performance in many crucial ways, and unless "main system management" is both effective and responsive, farmers' efforts to use water efficiently will not be fruitful (Wade and Chambers, 1980; Sundar, 1985). A distinction could be suggested between macro-management and micro-management, with farmers assigned the latter and agency managers the former. However it is not advisable to establish by definition such a dichotomous division of managerial labor. Optimum management is not so hierarchically organized with mutually exclusive spheres of activity.

It would be too large a task here to address the whole range of organizational and behavioral aspects of irrigation, including an analysis of irrigation bureaucracies. Assessing the possibilities for greater farmer participation in irrigation management is a substantial undertaking in itself. We proceed, however, with an appreciation that irrigation bureaucracies are important influences affecting farmers' performance, and we deal with the subject of user-agency relations in the concluding Chapter 9.

The reasons for considering a greater farmer role in irrigation management are reviewed in Chapter 2. There is no need for a lengthy justification since no blanket prescription of "farmer participation" is being offered. Questions about participation need to be answered in a disaggregated way, related to specific situations and objectives -- how much of what kind of participation, where and by whom is beneficial? Unfortunately, most treatments of the subject have been at the gross level of whether to have "farmer participation" or not. It should not be surprising if the decisions by administrators and engineers who have little personal experience and no conceptual preparation for dealing with the subject are not very precise or successful. They are like sociologists who with little agronomic training can only speak of soils as being "good" or "bad" or like political scientists who if called upon to design an irrigation system would probably overlook the provision of drainage facilities.

That decisions about organizational design get made with inadequate theoretical and empirical foundation is not attributable just to hubris. Technicians and administrators can argue that even if willing to invest more in the "software" rather than just in the "hardware" of irrigation, they could not get very satisfactory inputs from social scientists. There are many evident shortcomings in the social sciences when it comes to advising on irrigation. Too few social scientists have learned enough about soils or drainage to be useful collaborators with technical personnel. In the design, operation, monitoring and evaluation of irrigation systems, relevant social science contributions must be fitted to the physical conditions and constraints at hand. Prescriptions based on general principles are too abstract to be of much value.

Further, the divisions among disciplines have often led to social science analyses that were inadequate, e.g. sociologists not knowing enough economics to make useful recommendations, or anthropologists neglecting "power" considerations. Irrigation decision-makers need an integrated perspective that transcends disciplinary boundaries. The factors covered under the "socio" part of socio-technical analysis are not the preserve of any one of the social sciences. Fortunately, a new "field" of irrigation social science has begun to take shape in recent years (Coward, 1980).

For their work to be cumulative and reliably communicated, social scientists need some generally accepted concepts that are rigorous and relevant for irrigation management. They need analytical frameworks that make the varied and often amorphous tasks involved in irrigation more amenable to policy, planning and operational choices.

This study has three complementary objectives:

(1) to formulate and present <u>concepts</u> that can make more comprehensible the subjects of farmer organization and farmer participation for irrigation management, to aid practitioners as well as researchers and evaluators in understanding better certain possibilities and consequences;

(2) to assemble and assess <u>experience</u> with farmer organization and participation in irrigation management which can give empirical foundation to the subject, further illuminating possibilities and consequences;

(3) to derive <u>suggestions</u> for establishing farmer organization and participation which can contribute to improved irrigation performance.

Following a brief analysis in Chapter 2 of what farmer organization and participation can contribute to irrigation improvement, the first two objectives are addressed in Chapters 3 through 6. The third aim is covered in Chapters 7 through 9, extrapolating recommendations from the body of experience reviewed for this study. Before proceeding, however, readers should be introduced to the conceptual categories and data base which shaped this effort.

System Management by Agencies and by Users

Irrigation systems can vary in many respects: in their ability to adapt to changing conditions, in their productivity, their size, their complexity, their technology, their management structure. All of these facets interact, making analysis and prescription difficult. The "human processes of planning and management" referred to by Bottrall occur within irrigation systems that are created and operated:

(a) by some agency of government (or private enterprise),

(b) by the users of the system, that is, by the cultivators who utilize the irrigation water, or

(c) jointly by the users and an agency.[2]

In practice, there is a continuum between agency-managed and user-managed systems, with a middle range of joint management as the most common mode.

The polar alternatives of agency and user management are not equally feasible. Systems that are entirely agency-managed are for the most part hypothetical. Technical personnel can manage the higher levels of system operation by themselves on behalf of an outside agency, but rarely do they have the facilities, manpower and information to control and distribute water down to farmers' fields (Chambers, 1977). Some user role in water management is found therefore in virtually all systems, if only because it is prohibitively expensive for an irrigation department to try to handle all water management responsibilities itself. Even when the agency expects itself to deliver water to the turnout structures at the head of channels serving farmers' fields, in fact users are usually involved in some water management activities above the turnout, *de facto* if not *de jure*. The one exception we found in our review of irrigation cases was the Mwea scheme in Kenya, where the agency for some time delivered water right to the fields, and even ploughed them.[3]

While purely user-managed systems are more feasible and more common, they tend to be smaller in scale, previously designed and constructed by the users, and now operated and maintained by them. There are some instances of at least medium-scale user-managed systems, for example, the Chhatis Mauja scheme covering 7,500 acres in Nepal[4] and the zanjera systems in the Philippine province of Ilocos Norte which aggregate to approximately 42,000 acres. Some agency role in the management of such schemes may be introduced when they are "modernized" or rehabilitated with outside resources. When this occurs, they move into the broad middle range of jointly-managed irrigation systems. Even if an agency is responsible for operation and maintenance at higher levels of a system, users usually carry out some responsibilities in the lower reaches of system management.

In distinguishing between user-management and agency-management, there is no implication that socio-technical aspects can be divided between these two modes. The "social" dimension is not uniquely associated with users, as there are many important organizational and behavioral aspects of agency operation, nor are "technical" aspects the special sphere of agency personnel. Users often have considerable grasp of engineering, hydraulic and related principles even if they do not articulate them in formal scientific terms.[5] Agency and user management involves different sets of actors. How much each does of what, when and where has to be determined for particular cases and then to be assessed to see whether any changes could make for more beneficial water use.

Farmer Organization and Participation

Deficiencies in what Bottrall calls "the human process of planning and management" can occur at any and all levels of irrigation systems and in any of the roles, from supervising engineer to water user. We focus here on what can be done to improve farmer contributions to irrigation management, not because farmers' actions or lapses are regarded as the main "problem" but because there appear to be significant opportunities for improvement through working with water users in a more systematic way. While it is possible to have "too much" participation from the farmer side, practically all systems we looked at with an agency management role were well below any "optimum" level, so there appears to be considerable scope for productive increases in farmer participation.

Three basic propositions have emerged from the analysis:

(a) Farmer participation is not a single thing but a category encompassing many specific kinds of activities which water users can engage in; both theory and practice will benefit from distinguishing among and focusing on specific kinds of "participation."

(b) Farmer participation will be more predictable, productive and sustainable if it is channelled through organizations appropriate to the tasks of irrigation management.

(c) The physical nature of irrigation systems establishes different levels of operation, and the kinds of participation which are appropriate will vary according to where management activities occur in a system.

Farmer participation represents one part of the larger process of irrigation management, which as we have said is "socio-technical." It does not occur in an amorphous arena, but rather with reference to or through various structures. Some of these are physical, like the hydraulic structures referred to in (c) above, while others are institutional, reflecting established patterns of activity and norms. It is obvious that the channels and gates conveying and controlling water constitute a "structure" that shapes processes and outcomes. The same can be said of less concrete structures like water user associations, land tenure arrangements, or rules of water allocation.[6]

In this analysis, farmer organizations (structures) are seen as making water users' participation in water management activities (processes) more effective than if farmers were to attempt such activities on an individual basis. Social structures interact with and are affected by the physical structures which comprise what are usually referred to as "irrigation systems." Actually, organizational structures for management, whether bureaucratic or participatory, should be recognized as integral parts of such "systems," as both institutional and physical structures are intrinsic to socio-technical processes.

Farmer participation in irrigation management is multifaceted, dealing not only with water but also with the physical and social structures that control it. As analyzed in Chapter 3, irrigation management activities can be focused (a) on the water itself, (b) on the physical structures that capture, convey, distribute and remove water, or (c) on the social organizations that manage these physically-defined activities. These three sets of activities are interactive and interdependent, but it is useful to distinguish when the management effort is directed at the water, when it creates or controls the physical structures that control water, and when it affects the organization that manages both (a) and (b).

Organization should be understood in a functional way, not as anything rigid or abstract. In Chapter 3, four basic sets of activities are identified and analyzed which are particularly relevant to the tasks of water management, though they constitute the core of any organization:

(1) decision-making and planning,

(2) resource mobilization and management,

(3) communication and coordination, and

(4) conflict resolution.

This conception of "organization" makes it more tangible than do most definitions found in the literature, but it also makes the concept more operational. Two advantages of this approach are:

(a) Organization can be treated as a matter of degree. It exists to the extent that these activities occur on a regular and predictable basis. This gets around the formalistic and often meaningless question of whether or not there is "organization." In a functional sense there is almost always some organization, but it may not be very reliable or effective. One can assess the extent and effectiveness of organization by looking at who is performing what activities and how beneficially.

(b) Both formal and informal forms of organization can be more readily encompassed within a scheme of management. Organization is "formal" to the extent that these four activities occur according to explicit, written, possibly legal requirements. But there is still "organization" if they are based on implicit understandings and only social sanctions.

For irrigation planning and analysis, one should focus on the kinds and degrees of organization there are (or need to be) at different levels within an irrigation system. As discussed in Chapter 4, irrigation systems can be analyzed in terms of levels of operation that are hydrologically defined by the physical possibilities of controlling the flow of water. There are usually corresponding levels of organization. But the extent to which organization exists and is effective at a particular level is an empirical question. One can ask, for example:

> Is there formal or even informal organization for conflict resolution at the field channel level?
>
> To what extent are farmers involved on an organized basis in resource mobilization for channel maintenance above the turnout?
>
> Do farmers engage in regular decision-making about water allocation at the highest levels of the system, or even at lower levels?

As stated above, one should not expect the same kind of farmer participation at all levels, since water users' involvement in decision-making, resource mobilization, communication or conflict resolution will not be equally feasible or beneficial at all levels. Analysis of participation needs to be disaggregated by level as well as by the kind of activity in which water users could be involved. These four basic activities are ones in which agency personnel are also engaged -- with water users, separately from them, or even excluding them.

Empirical Bases for Improving Farmer Participation

Improving irrigation management requires better ways of thinking about farmer organization and participation, but there must be some foundation in actual experience to derive concepts that are rigorous and relevant, and recommendations that are practical. This study proceeded on two tracks, one conceptual and the other empirical. At the same time we were working to formulate a more coherent and useful framework for analyzing farmer organization and participation,[7] we were reviewing a set of 50 case studies compiled from the literature and analyzed comparatively according to a detailed protocol. These are listed in Table 1 at the end of this chapter (pages 14-17), together with the sources from which information on the cases was drawn.

These studies actually cover over 100 irrigation systems, ranging in area from ten acres to over a million. Denominating "case studies" is an impossible task. Lees (1973), for example, studied 24 irrigation systems in the Mexican state of Oaxaca for her very instructive monograph on irrigation management there. But she does not give comparative data on each case, rather describing the prevailing pattern of user water management and farmer-agency interaction. Martin and Yoder (1983), on the other hand, provide considerable data on the two hill irrigation systems they researched in Nepal, and in addition their research assistant has given comparable details on a nearby system with many similarities but some differences worth noting (U. Pradhan, 1982). We counted these Mexican and Nepal data sets each as one case study of farmer organization and participation, alongside, for example, the huge Office du Niger in Mali and the Daghara scheme in Iraq. Our purpose was to consider systematically a broad range of irrigation management experience rather than to construct a sample for purposes of statistical analysis. The precise number of "cases" thus is not so important.

As noted in Table 1, at the end of this chapter, some of the schemes are entirely user-managed and a few are basically agency-managed, with the majority exhibiting different kinds and degrees of joint management. Most are gravity schemes (storage or diversion), though some lift (pump) cases are included. Over half the cases are from Asia, where the greatest irrigated acreage is found (almost 90% of the area irrigated in LDCs). Cases were chosen principally for the adequacy of data they contained on farmer organization and participation in irrigation management, and secondarily for the variety of water management opportunities they presented, including geographic diversity. This latter consideration meant that statistically speaking, African, Latin American and Middle Eastern cases are "overrepresented." But this seemed justifiable to gain a better overview of irrigation management experience worldwide. We believe the data base informing this analysis is reasonably representative. Certainly it is very rich in variety and insights.

The analysis offered is a state-of-the-art exercise. It cannot be a definitive study because the knowledge base is rapidly expanding, as seen from the dates of most of the publications listed in the Bibliography. Two-thirds of the materials have appeared in the last five years, and from the accelerating rate of publications, we may expect a doubling of the knowledge base in the next five years.

This makes it all the more important to have analytical frameworks that can help make assessments of experience more comparable and cumulative. We have tried to present concepts and terms that are theoretically sound and practically useful. They are meant to be of utility to researchers and practitioners alike, so that knowledge can more easily pass back and forth between the academic and policy realms.

NOTES

[1]One might ask, why not use the reverse designation -- "techno-social"? But this seems less satisfactory, perhaps implying that social aspects are more basic.

[2]This distinction made by Coward (1980a:27) parallels the classification of Chambers (1977), who proposed distinguishing among systems where the acquisition and allocation of water is (a) by the bureaucracy, (b) by the community, or (c) by a combination of the two (bureaucratic-communal). It is desirable to avoid the often prejorative term "bureaucracy," and the designation "community" implies that whole communities are involved in irrigation management, which is seldom correct.

[3]References for cases mentioned in the text are given in the following listing to avoid having to repeat the references each time a case is referred to. Full citations are given in the References on pages 177-199. In the Kenyan case cited here, this overextension of agency responsibility has subsequently been reduced.

[4]The Chhatis Mauja scheme has a three-tiered structure of organization, which even links to an informal fourth tier that includes three other user-managed systems receiving water from the same river source. So user-managed schemes need not be limited to "small" systems. The four systems irrigate over 25,000 acres without any agency involvement.

[5]In Chapter 3, cases from Mexico, Nepal and the Philippines are cited where farmers proved better able to make predictions about the feasibility of certain dam designs than did the project engineers. If farmers can be correct about straightforward matters like dam design, their views on technical subjects should not be excluded by assigning user activities to the "social" sphere.

[6]We will not go into theoretical distinctions between "institutions" and "organizations," which are analyzed in Uphoff (1986a), especially Section 1.5.

[7]This work was facilitated by previous work done at Cornell under the auspices of the Rural Development Committee on farmer organization and participation which included Uphoff and Esman (1974), Cohen and Uphoff (1977), Uphoff, Cohen and Goldsmith (1979:213-234), and Esman and Uphoff (1984).

Table 1: CASES ANALYZED

ASIA

SOUTH ASIA	COMMAND AREA (in acres)	MANAGEMENT	SOURCES
India			
1. Pochampad (A.P.)	60,000	Agency + users (organized by agency)	Singh (1982, 1983, 1984)
2. MNC-TNC (A.P.) (24 communities)	400,000	Agency + users (organized by users)	Wade (1979, 1984a)
3. Ahar-Pyne (Bihar)	100s	Users (by informal organization)	Sengupta (1980, 1984)
4. Sone (Bihar)	1,450,000	Agency + users (organized by agency)	Pant and Verma (1983)
5. Bhakra (Haryana)	4,000,000	Agency (warabandi water rotation)	Vander Velde (1980), Reidinger (1974)
6. Sananeri (Tamil Nadu)	440	Users (agency at higher level)	Meinzen-Dick (1984)
Pakistan			
7. Upper Bari Doab (Punjab)	1,000,000	Agency (warabandi water rotation)	Lowdermilk et al. (1975)
8. Chaj Doab (Punjab)	1,680,000	Agency (warabandi water rotation) (note: both Punjab cases are supplied by the same large Indus River system)	Merrey (1982, 1983, 1983a 1984)
9. Daudzai (N.W.F.P.)	35,000	Users (liaison with agency management)	Bhatty (1979)
Sri Lanka			
10. Pul Eliya	132	Users (indigenous organization)	Leach (1961)
11. Minipe	15,000	Agency + users (organized by agency)	de Silva (1981, 1984)
12. Gal Oya (Left Bank)	60,000	Agency + users (organized by agency)	Abeyratne et al. (1984), Uphoff (1982, 1985, 1986), Widanapathirana (1984), Wijayaratna (1984)
Nepal			
13. Argali, Chherlung and Tallo Kulo	116, 085 and 34	Users (variety of indigenous organizations, liaison with agency)	Martin and Yoder (1983), U. Pradhan (1982), Martin (1986), Yoder (1986)
14. Chhatis Mauja	7,500	Users (indigenous organization)	P. Pradhan (1983, 1984)
Bangladesh			
15. Pultan Para	75	Users (liaison with agency)	Howes (1984)

SOUTHEAST ASIA				
Philippines				
16.	Bacarra-Vintar (Zanjera)	1,260	Users (federated organizations)	Lewis (1971), Siy (1982)
17.	Lalo and Baris National Systems	7,000 and 5,440	Agency + users (organized by agency)	Illo and Nestor (1981), Illo and Chiong-Javier (1983)
18.	Aslong and Taisan Communal Systems	1,250 and 425	Users + agency	Inos (1981)
Indonesia				
19.	Tihingan (Subaks) (Bali)	575	Users (indigenous organization)	Geertz (1967), Birkelbach (1973)
20.	Pakalen Sampaen (Java)	685,000	Agency + users (rotations)	Taylor and Pasandaran (1979)
21.	Bima and Tayuban (Dharma Tirta) (Java)	395 and 1,030	Agency + users (organized by agency) (note: these are part of larger water supply systems having several sources)	Duewel (1982, 1984), Adams (1983)
Thailand				
22.	Seraphi and Sankaemphaeng	2,500 and 4,000	Users	Potter (1976), Abha (1979),
23.	Nong Wai	25,000	Agency and Users (organized by agency)	Kathpalia (1984)
Malaysia				
24.	Muda	200,000	Agency + users (organized by agency)	Afifuddin (1978)
Laos				
25.	Nam Tan	2,600	Agency + users (traditional roles)	Coward (1976)
Papua-New Guinea				
26.	Wamira	1,250	Users (agency technical intervention affects traditional organization)	Kahn (1983)
EAST ASIA				
China				
27.	Meichuan	20,000	Agency + users	Nickum (1981)
Taiwan				
28.	Chang Hua	15,000	Users (Irrigation Assoc.) + agency	Stavis (1982)
29.	Namton, Taoyuan, Yunlin and Chianan	33,000 to 267,000	Users (Irrigation Assoc.) + agency	Abel (1975), Moore (1983)
South Korea				
30.	"S. Y." F.L.I.A.	26,500	Agency + users (organized in Farm Land Improvement Association)	Wade (1982b)

MIDDLE EAST

Iran			
31. Deh Salm and Nayband	70 and 60	Users (quanat system)	Spooner (1974)
Iraq			
32. Daghara	250,000	Agency + users (tribal associations)	Fernea (1970)
Oman			
33. Izki	200	Users (quanat system)	Wilkinson (1977), Sutton (1984)
Egypt			
34. Abu Raha	200,000	Agency (the name is that of the village area within a much larger system; no user organization)	CSU/CID (1980) and other project publications

AFRICA

Senegal			
35. Bakel (many systems)	10 to 15	Agency + users	Adams (1977)
36. Matam (many systems)	37 to 62	Users + agency	Fresson (1979), Diemer and van der Laan (1983), Patterson (1984)
Mali			
37. Office du Niger	125,000	Agency	de Wilde (1967)
Ghana			
38. Tono	6,000	Agency	Chambas (1980)
Zimbabwe			
39. Sabi River (9 schemes)	112 to 928	Agency	Roder (1965)
Tanzania			
40. Sonjo	500	Users (indigenous organization)	Gray (1963)
Kenya			
41. Marakwet (many systems)	21 to 384	Users (indigenous organization)	Ssenyongo (1983)
42. Mwea	15,000	Agency (some user organization introduced)	Chambers and Moris (1973)
Sudan			
43. Gezira	1,850,000	Agency (some user organization introduced)	Gaitskell (1959), Simpson (1976), Bailey et al. (1981), Elder (1982)
44. Rahad	315,000	Agency + user (organized by agency)	Benedict et al. (1981)

LATIN AMERICA

Brazil			
45. Morada Nova,	2,875	Agency	Hall (1978)
San Goncalo,	1,500	Agency	
and Sume	650	Agency	
Chile			
46. San Pedro de Atacama	3,750	Users + agency	Lynch (1978)
Peru			
47. Quinua (Lurin Sayoc and Hanan Sayoc)	2,000	Users	Mitchell (1976)
Ecuador			
48. Quimiag (Huerta Redonda)	1,500	Users	Cornick (1983)
Mexico			
49. Zapotec (Diaz Ordaz)	375	Users	Downing (1974)
50. Oaxaca (24 systems)	3,750 (ave.)	Users + agency (latter role is undermining users' organization)	Lees (1973)

2
Why Farmer Participation?
Contributions to Irrigation Management

Farmer participation is presently more often a potential than a reality. In agency-managed systems, the users' role is usually restricted to activities "below the outlet," discussed in Chapter 4. Even in user-managed systems, the amount and effectiveness of participation can be less than desirable, for example, if decision-making is dominated by rural elites, or if water distribution does not reach all tail-end farmers. It is important to be clear about the goals of irrigation management, to have some criteria by which to judge when more or less participation may be desirable, and also how much and what kinds? Furthermore, the benefits and costs need to be considered, preferably in relation to one another, though there are few analyses which permit such comparison.

Objectives in Irrigation Management

The goals that may be furthered by farmer participation in irrigation systems management are seldom of equal importance to all concerned. The things sought by farmers and by government agencies may or may not be congruent. Moreover, farmers may disagree among themselves on the weighting of objectives, as may government agencies, according to their responsibilities and interests. Some goals, fortunately, may be similarly appreciated by farmers and officials.

The objectives themselves are often interrelated and cannot be ranked according to some fixed priority. From the literature we can identify the following criteria, some instrumental to others. The first and last categories represent the broadest goals, whereas the others are more specific and sometimes intermediate. All can be furthered by farmer participation in various ways.

A. GREATER PRODUCTION OR PRODUCTIVITY, measured either as total output or as the amount produced per acre or per unit of water, to be achieved through some combination of increases in:

1. **Yield,**

2. **Area cultivated,** and/or

3. **Cropping intensity,** i.e. more crops produced in a year.

Such increases can come from more adequate or more timely water application, from savings of water within or between seasons, or from use of new technologies, such as higher-yielding varieties or fertilizer and agricultural chemicals, made possible or more profitable by better water management.

B. IMPROVED WATER DISTRIBUTION, which has two aspects:

1. Greater **reliability** and **predictability** in the amount and timing of water deliveries; this can encourage use of new technology and make possible more efficient use of labor.

2. Greater **equity** of distribution, particularly between upstream and downstream areas; this can contribute also to production and productivity.[1]

C. REDUCTIONS IN CONFLICT:

1. Among users, e.g. between upstream and downstream farmers, so that cooperative water use is more possible.

2. With government agencies, so that main system management is less impaired or less subject to political interference.

D. GREATER RESOURCE MOBILIZATION:

1. Contributions of **labor and materials:**

a. during **construction** or **rehabilitation,**

b. for system **operation,** at lower levels and upwards as established by agreement or precedent, and

c. for system **maintenance** (routine, preventive and/or emergency), at lower levels and above.

2. Contributions of **funds:**

a. toward the **capital costs** of construction, and

b. toward the **recurrent costs** of operation and maintenance.

3. Contributions of **information:**

a. during **design** activities or rehabilitation planning, and

b. for improved **operation and maintenance.**

4. **Cost and quality control** by farmers such as inspection of work by contractors or agencies, scrutiny of materials delivered for use, and overseeing work at field level.[2]

E. SUSTAINED SYSTEM PERFORMANCE:

1. **Managing soil and water resources** so that their productivity is maintained; for example, too little or too much water application can lead to salinity, depending on soil and water table conditions.

2. **Achieving and continuing intensified production** so that a larger population base can be supported.[3]

Production objectives probably command the broadest support among both farmers and government agencies, though governments tend to look at aggregate levels of production. Farmers, on the other hand, are more concerned with <u>who</u> is getting the production increase and with what net benefit. The latter is affected by the prices farmers receive for their outputs and have to pay for their inputs. What looks like satisfactory irrigation system performance to the government, in terms of gross physical production, may be judged unacceptable to farmers, in terms of their resulting income.[4]

What commodities will be produced on irrigated land may be a matter of contention between farmers and officials, as the latter may prefer and promote cash crops which farmers do not find profitable or which involve more risk or effort than farmers judge worthwhile.[5] Or farmers may prefer high value crops, which require more water than approved by the government for that area (for example, in India, where cropping "zones" are established). Greater scope for farmer participation will tend to lead to production of crops which farmers prefer. This can be at variance with what agencies want to promote.

Improved water distribution is welcomed in principle by farmers as well as officials, though farmers will usually emphasize reliable and predictable deliveries with upstream water users not necessarily wanting equitable distribution if it means reallocating water away from them. Governments endorse equity in water distribution but often find promoting it difficult, because their personnel may not be able or inclined to achieve it. We have been impressed in our literature review by how often farmers have devised methods -- organizational or technical -- for equitable water distribution, suggesting a collective value placed on this objective. However, consensus is not always reached in specific instances where interests come into conflict and definitions of "equity" can vary considerably from place to place, and among users.

Conflict reduction is often of special concern to farmers, though agencies also have an interest in promoting it, if only to diminish the problems it causes for them. It is hard to know how much value is attached to this objective because of difficulties in measurement. We have found in Sri Lanka that reduced conflict appears to be one of the payoffs which induces farmers to make special collective efforts to achieve a more equitable distribution of water.[6] A low level of conflict may mean either that there are few sources of strife in the situation or that existing institutions, formal and informal, are capable of controlling it. In the next chapter we will be considering conflict resolution as a generic activity invariably required for irrigation operation.

Governments often equate farmer participation with **resource mobilization.** Where an irrigation agency must "pay its own way," resource mobilization is likely to be the objective with which the agency is most concerned. The innovative efforts of the National Irrigation Administration in the Philippines to help farmers organize and to participate in design and construction activities were prompted in part by a policy decision to recover from users the capital costs of improvements (F. Korten, 1982). Farmers, for their part, may regard the possibility of mobilizing government resources as one of the reasons for them to organize. Farmers' enthusiasm for resource mobilization will be conditioned by the extent to which the contributions required from them yield compensating benefits such as better water supply and distribution. To the extent that farmers' contributions of knowledge are solicited in the planning or operation of systems, other contributions will be more readily forthcoming from them (Coward and Uphoff, 1985).

The goal of **sustaining system performance** should be generally agreed upon, but because it represents a deferred benefit, it is often overlooked. It is not clear whether farmers or officials tend to be more interested in this issue. There are several cases in our study which show declining production and capacity to support the population. Yields are reported as already declining in the relatively new Rahad scheme in the Sudan, while they have been stagnant for the last 20 years in the famous Gezira scheme in the same country. In some instances a decline is attributable to inadequate water control or neglect of drainage activities, resulting in waterlogging. It can also be due to a lack of incentives to invest in maintaining fertility.[7] We expect that sustaining stable and productive irrigation systems is going to become a much more salient objective of system performance in the years ahead.

This last criterion highlights the possibility that efforts to increase production in the short-run can undermine capacity for long-term productivity. Installing too many tubewells, for instance, can draw down the water table over time, or excessive watering for short-run gains can result in salinization of the soil. Trying to achieve very high levels of resource mobilization from water users can similarly lead to conflict which reduces organizational capacity to achieve other goals.

It should be clear that there is no necessary symmetry between the goals of farmers and those of government agencies that engage in irrigation management. Such conflicts over objectives have been documented in a Philippine case by Svendsen (1983). If government goals include attempts to maximize production while maintaining an artificially low price, for example, to keep consumers satisfied, the farmers whose crops are not very profitable will have little interest in "efficient" use of water which means more poorly remunerated work for them. Under such circumstances they will want to reduce weeding or to facilitate land preparation by irrigating "excessively" (in the opinion of officials).

Farmer participation will be more evidently useful and more sustainable to the extent that it is contributing to the achievement of objectives which farmers themselves value. Where the government seeks farmers' cooperation in activities whose outcome it values more than they do, it must be prepared to provide some compensating benefits which farmers appreciate, or to use some form of coercion. But such measures may not be successful, as seen in the Sabi Valley schemes in Southern Rhodesia (now Zimbabwe) and in the Rahad project in Sudan. Congruence of objectives between water users and system managers is among the most important features contributing to productivity as well as to farmer participation in jointly-managed systems.[8]

Benefits of Farmer Participation

The literature gives only fragmentary evidence of precise gains from farmer participation in irrigation management, though it is replete with descriptions of benefits attributable to participation. Examples would include the report of a 30 percent increase in the flow of water to the downstream half of the Minipe scheme in Sri Lanka within the first year of introducing farmer organizations there. In the Pochampad scheme in India, the irrigable area was extended by 25 to 35 percent due to the rotational system which came into operation after Pipe Committees were established. With the help of these committees, the time required for land development of a turnout area (chak) was reduced from one year to only 4 to 6 months. In the Nong Wai scheme in Thailand, farmer organizations reportedly raised cropping intensity from 50 to 90 percent in two years' time. When a "participatory" approach was taken to expanding the Buhi-Lalo scheme in the Philippines, engineers with farmer advice and concurrence could reduce the planned length of field channels by 12 percent, thereby saving substantial costs. The construction work done by the farmers was completed four months ahead of schedule, and project engineers judged the quality of the work to be better than average (Illo and Chiong-Javier, 1983:xxi-xxv).

Less precise but hardly less significant gains are reported from the Muda irrigation scheme in Malaysia. When it was first opened, there was "anarchy," according to Afifuddin (1978). Within several years this situation was replaced by some degree of order through the establishment of farmer organizations, which produced noticeable improvements in economic and social performance. In aggregated terms, Lowdermilk (1985:2) reports that farmers contributed $7.6 million worth of labor in a large ($42 million) program to rehabilitate turnout areas (chaks) in Pakistan. It is estimated that users are providing 30 percent of the cost for a World Bank rehabilitation project in Pakistan (Fairchild, 1985).

One can look at the benefits of farmer participation conversely, by considering the difficulties or costs where irrigation projects are operated without user inputs to management, even at lower levels of the system. The Gezira scheme in the Sudan was one of the first major agency-planned and operated systems, also one of the largest, over 1.8 million acres. Its early economic success encouraged other countries to embark on similar regimented, large-scale schemes such as the Mokwa project in Nigeria (Baldwin, 1957). Unfortunately, crop yields in Gezira have been stagnant for the last twenty years, and the case materials suggest that this is due more to social and organizational factors than to technical constraints. Tenants -- as farmers in the Gezira scheme are called -- do not see themselves as "partners" and thus have not been responsive to opportunities for innovation, according to the case documentation.[9] To achieve more personal commitment and attachment to the project, its managers have now granted the Tenants' Union a voice in running the scheme, but the scope for farmer participation is still too limited to give them much incentive for change. Poor agricultural performance in the Egyptian case of Abu Raya was similarly attributed in part to the lack of self-respect and self-confidence possessed by farmers who have no active role in irrigation management. This suggests certain economic costs of not having farmer participation, but such costs are very difficult to measure.

One kind of farmer participation often overlooked and not encouraged is involving users in planning and operations. The case materials document that farmers have not only social skills for problem-solving but also valuable technical knowledge about acquiring and controlling water, as noted in Chapter 1. In China, irrigation authorities in Meichuan accepted a "melons-on-the-vine" strategy after seeing a farmer-built system with many small reservoirs storing and controlling water along a main channel. With farmer help, they constructed 21 small reservoirs and over 6,000 ponds to add almost 30 million cubic meters of storage capacity to the 27 million cubic meter capacity of the main Meichuan reservoir.

In irrigation schemes in the hills of Nepal, tunnels have been constructed through portions of mountains where the channels installed along the sides of those mountains have been frequently damaged by landslides (Martin and Yoder, 1983). This requires considerable skill in design as well as construction.[10] In Kenya, Marakwet farmers have developed a furrow system taking water from the Rift Valley escarpment by ingenious channels to cultivable areas up to 9 miles away.[11]

While such techniques may not apply to large-scale schemes, they testify to the empirical knowledge which farmers can have of hydrology and engineering. Moreover, since most of the major opportunities for large-scale irrigation development have been identified and developed, additional area is likely to be added in less obvious and less favorable circumstances, that require more inventive design and which could benefit from the low-cost construction techniques that farmers may know or be able to devise.

There is no guarantee that water users will always know or be able to apply appropriate technologies for irrigation. A case in point from our literature review is one where the Bangladesh Agricultural Development Corporation (BADC) installed a deep tubewell at Pultan Para. It was up to the group of users to lay out and build a system of major and minor field channels, which they did rather badly. In such a situation, more technical advice from BADC would have been helpful for making the best use of available water. The extent of farmer competence to contribute to solving technical problems is something to be examined rather than assumed. Presently it is too often assumed that users have nothing to contribute technically.

Costs of Farmer Participation

The Bangladesh example just discussed suggests the possible negative side of farmer participation. Also, we know that practically anything which has benefits is likely to have some associated costs, and participation is no exception. This suggests that participation should be "optimized" rather than maximized. Practically all kinds of participation, discussed in the next chapter, have some costs to farmers, if only in terms of time, whether representing forgone earnings or leisure. A participatory system of water distribution which enforces equitable division between head and tail areas may take away some of the advantages enjoyed by head-end farmers, which is a cost to those users. To agency managers of irrigation systems, a greater farmer role may be seen as reducing their authority or status (or even income gained illicitly). Some of these costs may not be judged worth taking into account, but they will affect the incentives farmers and officials have for cooperating in a more participatory system of irrigation management.

In our review of the literature, we found many more comments on positive effects of participation than negative ones, though this may reflect some bias in the literature.[12] Several cases were reported in the previous section where there were costs of <u>not</u> having farmer participation. We found that most of the time where changes were introduced by farmers or by agencies, it was in the direction of getting more rather than less user participation. Occasionally we found governments reducing the scope of farmer participation as done initially in the Sederhana project in Indonesia. In this case, when communal irrigation schemes were rehabilitated by an agency, it took over responsibilities that farmers had previously handled themselves. As this was costly and unnecessary, the government began experimenting with programs similar to those in the Philippines and Sri Lanka, discussed below, employing organizers to promote farmer participation in rehabilitating and then operating and maintaining the small-scale systems (Morfit, 1983; Robinson, 1985).

Even some schemes controlled by government agencies have moved, however haltingly, toward providing for more farmer participation. The Muda case in Malaysia and the Gezira project in Sudan were mentioned previously. In the Mwea scheme in Kenya, agency managers have sought to introduce some organizational channels for farmer participation, even if only as "a useful safety valve" (Chambers and Moris, 1973:316). Case materials suggest that a larger role for farmers would have been useful not just in the Tenants' Liaison Council and the Coop Credit Union but also in irrigation matters as well.

Comparisons of Benefits and Costs

Where farmer participation has developed spontaneously or is simply encouraged without any agency efforts to promote it, there are no costs or they are likely to be small. Agency personnel may feel that they are losing something in terms of status and power. But economic gains from the system can offset this, and even give enough satisfaction and credit that staff accept this new arrangement, as discussed in Chapter 9. There may be cases where the efficiency of water use and/or the equity of water distribution is so satisfactory that it would decline if water users had greater responsibility for management. But such well-run systems are definitely a minority and probably have already developed satisfactory mechanisms for farmer involvement. Judgments must be made in each case about how close an irrigation system is to some optimum with respect to the criteria discussed at the beginning of this chapter, and whether performance levels could be increased through farmer participation.

Where irrigation systems are found wanting and a greater role for water users appears useful, the question becomes how to establish this. There have been a number of efforts made in various countries, discussed in the following chapters, to introduce farmer organization, commonly called water user associations (WUAs), to achieve more farmer participation in irrigation management. It can reasonably be asked whether such efforts are cost-effective, whether the benefits therefrom exceed the costs.

This issue is difficult to answer definitively since there are many intangible benefits and costs that elude any summary measure comparing the two. Such factors should be taken into account, but policy-makers and planners usually want to know at least the ratio of benefits and costs that can be denominated in monetary terms to see what the narrowly economic impact of such a program would be. While such measures are likely to be only partial reflections of what is accomplished for a given expenditure, they should be estimated, accompanied by whatever qualifications and additional calculations seem warranted.

We have been able to find in the literature only two systematic comparisons of benefits and costs where farmer organization has been introduced to improve irrigation management. As seen above, there are various reports of benefits _or_ costs, but seldom are the two compared for the same program. These two estimates, from the Philippines and Sri Lanka, are very encouraging, suggesting economic rates of return in the range of _50 percent_. This is not considering intangible benefits, which probably exceed any unaccounted costs by an even greater margin. Such a rate of return on "software" is several times greater than that accepted now for investments in irrigation "hardware."

In an analysis of benefits and costs in 19 pilot irrigation schemes where the participatory approach was followed in the Philippines, direct quantifiable benefits (savings) in the construction phase were $24 per hectare, against a cost for the community organizer program (salaries, training, etc.) which came to $49 per hectare. This left a negative balance of $25 per hectare in direct costs at the time of completing rehabilitation. (Even this cost could be seen as offset by the fact that field ditches, which cost $38 per hectare to build, remained intact and were used, thanks to farmer consultation and involvement, whereas in many other schemes they were torn up within a few years; but this was not included within the benefit-cost assessment.) Since the value of farmer resources mobilized for operation and maintenance was calculated to be $12 per hectare annually, in narrow financial terms, the cost of the program could be "recovered" within some two years, and the stream of benefits should continue thereafter without additional or with little investment (Bagadion and F. Korten, 1985).

Where farmer participation was introduced in a large irrigation scheme in Sri Lanka (Gal Oya), definite net benefits could be seen within two years. In a pilot area of over 10,000 acres, where organizers had been fielded to help farmers establish water user associations, the cost of the program including all training, supervision and salaries, was about 60 rupees per acre per season. Direct benefits from increased production came to about 90 rupees per acre per season, figuring only the value of maintenance work done by farmers and of increased production from just one tail-end area that remained uncultivated before farmers worked out and implemented a system of rotational water distribution.[13] Now that the program has been established and can move into a "maintenance" phase with less intensive support from organizers, the ongoing cost including supervision, transportation, on-going training, etc., comes to about $1 per acre per season. This is more than justified if improved farmer operation and maintenance can raise production by even 1-2 bushels per season (2-4 percent), an easy target to meet.

Additional benefits not included in Wijayaratna's calculations because of lack of data or difficulties in quantification, were (a) reduced damage to physical structures by farmers and animals, (b) reduced conflicts over water,[14] and (c) yield increases attributable to more reliable water distribution at the field level which encourage adoption of new technology. By calculating the marginal economic value of irrigation water, benefits could be attributed to increased efficiency in water use; alternatively one could value the production from the additional area that could be cultivated due to more sparing use of water upstream.[15]

Still more difficult to measure is the value of improved system performance indicated by marked reductions in the number of irrigation-related complaints since farmer organizations were established.[16] This represents a level of satisfaction that can translate into social and political benefits. Thanks to physical rehabilitation of the Gal Oya system, the facilities have been improved, as has the performance of the Irrigation Department, partly in response to being able to (or having to) work with organized water users. The main reason for fewer irrigation complaints being brought to senior politicians and officials has been the greater cooperation among farmers, who could solve many of the problems by themselves once organized, and cooperation between farmers and government staff, who have become more engaged in problem-solving with the result that higher levels are less often bothered.[17]

The Philippine and Sri Lanka programs have encountered many difficulties and both have fallen short of their own goals in many ways, so they are not perfect "models" to be replicated. As discussed in Part III, many positive lessons can be learned from their experiences, however, and from the "learning process" approach adopted in both. Programs for introducing water user associations into schemes of irrigation management could be more successful or less successful than these (indicated in part by higher or by lower benefit-cost ratios).

The relevant consideration is that agencies bearing the cost of establishing organized farmer participation have realized significant returns from such investments, in the range of 50% according to these cases, considering only measurable and tangible benefits. This should encourage government and donor agencies to look seriously at the possibilities for trying to improve irrigation management by involving farmers more systematically. The analysis and experience reported in Part II and the suggestions offered in Part III should aid in analyzing situations and formulating programs of action.

NOTES

[1]Improvements in distribution can contribute not only to equity, thereby creating a valued sense of well-being and of fairness in the community, but redistribution will increase total production if output gains downstream are greater than any output declines upstream. (In some cases, reduction in excessive water offtakes might even increase upstream yields.) Other benefits from equitable distribution include possibly having more time for other purposes (not having to spend time defending or attacking unequal distribution).

[2]This latter function is not common but was undertaken by water user associations in the Philippines (D. Korten, 1980; F. Korten, 1982). Quality and Quantity Control Committees were set up during construction to oversee work and materials at the construction sites for new diversion dams. It was in farmers' interest to insure that no inferior materials or short deliveries were accepted, because they had agreed to repay the capital cost of permanent structures and better canal systems. They were already reducing the amount of their financial obligation by contributing labor and materials to the construction effort.

[3]This is not an "ideal" objective but one which should be kept in mind, particularly in some of the densely-populated parts of Asia. Chambers (1977) refers to this objective as "carrying capacity."

[4]In two of the three schemes reported in the Brazil study, for example, a majority of farmers were operating at a loss and going more deeply into debt (Hall, 1978).

[5]This conflict appears most frequently in large-scale resettlement schemes, particularly in Africa, where the main purpose of the scheme (from a government point of view) is cash crop increases. Strict cropping and production controls were found, for example, in Egypt (Abu Raya), Sudan (Gezira and Rahad), and Kenya (Mwea). This was also true in some smaller schemes such as the requirement of tobacco production in Indonesia (Bima and Tayuban). Farmers may be more interested in growing subsistence crops or a crop not subject to price or marketing controls.

[6]The introduction of water rotating and water saving measures in the Gal Oya irrigation scheme to help water-short "tail-enders" is described in Uphoff (1985). Singh (1984) also reports that some of the cooperative behavior of farmers in the Pochampad scheme in India could be attributed to their desire to reduce conflict. This enhances a sense of group solidarity which can help members cope with other problems in time of need.

[7]Farmers in the San Pedro de Atacama area of Chile have reverted largely to subsistence production partly because salinization has reduced their productive area but also because decommercialization eases their problems of water management now that control over supply has been lost to upstream users. When irrigation was better managed -- in the time of the Incas -- the population supported by the area was much greater than at present (Lynch, 1978).

Merrey (1983) reports that in his Pakistan case, "Gondalpur residents are not more prosperous now than they were before they became the 'beneficiaries" of irrigation. Per capita production and consumption have actually decreased because population growth has not been matched by concomitant increases in production. The extraction of resources from farmers through state taxes, indebtedness and other means has left them with little margin left for investment in better technology or water control." This area is also afflicted by waterlogging which has reduced the cropping intensity in recent decades.

[8]There is little consistency in the terminology and criteria for evaluating irrigation system performance, so we have offered a

scheme in Annex 1 (pages 161-164) that organizes criteria for assessing water supply.

[9]As far back as 1966, a World Bank team suggested more freedom of choice for tenants in what they grow and how they manage their on-farm operations.

[10]An impressive indigenous technology comparable to that in Nepal are the intricate underground tunnel networks known as qanats conveying water to irrigation systems found throughout the Middle East, from Pakistan to Morocco. Our literature review included qanat cases from Iran (Spooner, 1974) and the Persian Gulf state of Oman (Wilkinson, 1977). In the latter case, construction of an inverted siphon across a wadi was reported. These systems, however, are built by specialists and financed by outsiders, so they demonstrate a high level of indigenous but not necessarily farmer technical skill.

[11]Ssennyonga (1983) writes of technology used already two hundred years ago:

> ... there are spots where natural phenomena such as deep gullies, jutting rocks or stones made it impossible for water to flow to certain areas by gravity propulsion. In these places, ancillary structures had to be erected. In some places, water canals are suspended up to 15 feet above the ground. The construction of these pole-supported structures demanded considerable ingenuity; for example, the constructors had in some cases to be suspended by a network of ropes manually held by a team of strong men. In other instances two or even three water furrows flowing in different directions had, due again to physical barriers, to pass through one narrow point. In such cases, wooden aqueducts (dug-out tree trunks) were used to enable at times several furrows to flow on top of one another.

When observing the Marakwet man-made furrows -- 2 to 3 feet in width -- flowing on the ground, one would hardly know they are not naturally-occurring streams.

[12]There has been some tendency in what gets reported to look at participation uncritically. One of the few studies attempting to assess the effects of farmer participation empirically (Robinson 1982) found little difference in yield or payment of fees

associated with different degrees of organizational effectiveness in two large Philippine schemes. Unfortunately, to get comparability across a large number of cases, he relied more on interview data than on direct data, and the cases did not necessarily represent the full range of variation. No quantification and summation of costs and benefits was made so as to arrive at an assessment of net costs or net benefits. Robinson did document quite different patterns of participation and performance between wet and dry seasons, something often ignored in plans and evaluations.

[13]The water saved by rotations in the M5 sub-system permitted cropping of 717 acres (Wijayaratna, 1984). Production data gathered by the Agrarian Research and Training Institute, which was responsible for introducing the farmer organizations, showed a net profit from paddy production in that season to be 23 rupees per bushel, so this figure was used to calculate the value of added production (rather than the gross sale price). The training costs of the program might reasonably have been "capitalized" over more than two years, which would have improved the benefit-cost ratio, but there was enough turnover of organizers (who were given only temporary appointments) that training was figured as a current expense.

[14]Both farmers and Irrigation Department officials agreed these had declined. The farmer chairman for M5 sub-system said that before the program there were even murders over water in his area, but now there were hardly even conflicts any more (personal communication).

[15]With physical rehabilitation and with cooperation from farmer organizations, water issues for the 1984-85 and 1985-86 main seasons were brought down to 2 acre-feet per acre, less than the national norm of 3 acre-feet and much less than previous issues for the Left Bank system. For the 1985 dry season, the water issue was reduced by one-third, bringing it down to the national norm (seldom achieved). The end-of-project evaluation calculated an internal rate of return of 47 percent, due mainly to an extension of irrigated area (ISTI, 1985). Some of the assumptions on which this figure was based might be questioned, but the project was reasonably surely a profitable one. Economic benefits were attributed mostly to the investment in physical rehabilitation (which absorbed most of the cost -- the "software" investment in farmer organization was less than 5 percent). But getting water savings to serve an enlarged command area requires farmer cooperation which had been previously lacking, both in managing water supplied to the

field channels fairly and effectively and in not sabotaging distributions at higher levels.

[16]The District Minister has stated publicly that five years ago, out of every ten farmers he met, eight would have complaints about some irrigation problem; but now nobody comes to him about irrigation. The Government Agent, the chief administrator officer in the district, said in a published interview that when he came in 1980, on his days to meet the public, he would have a hundred people lined up to complain about irrigation problems; now "not a single person comes." Desatiya (Sinhala), October, 1984, p. 19. The Deputy Director of Irrigation in charge of Gal Oya area reports that the number of registered letters he receives with complaints about irrigation (with copies to the Minister, Prime Minister, etc. which take a long time replying to) has dropped from hundreds every season to very few (S. Sentinathan, personal communication). With good linkages between farmers and irrigation officials, the role of politicians in water matters has almost disappeared.

[17]These results are more credible because similar outcomes occured when the Deputy Director for Irrigation responsible for the 15,000 acre Minipe scheme near Kandy experimented with introduction of a system of farmer organization and participation there (de Silva, 1981 and 1984). This engineer is now Chairman of the Mahaweli Engineering and Construction Authority in Sri Lanka.

Part Two

Analyzing Farmer Organization and Participation

3
What Kinds of Participation?
Activities in Irrigation Management

The interaction of physical and organizational aspects of irrigation makes it a socio-economic process, exemplified by the three focuses of irrigation activity which are closely linked with one another:

* Some activities focus on the water which is to be provided in an adequate and timely manner to crops;

* Other activities focus on the structures which give control over the water for its application to crops; and

* Still other activities maintain the organization of effort which can manage the structures that control the water.

It is probably not coincidental that these three focuses correspond to the three factors of production which economists classify as: (1) land, the term used for all natural resources; (2) capital, created from other resources to make them more productive; and (3) labor, covering all human activity. Water is a crucial natural resource, generally renewable within some limits. The physical structures for irrigation, like other kinds of capital, are produced through investments of materials and labor. Organization is established and maintained through human efforts, embodying both energy and ideas, which may come from users, from agency personnel, or from some combination of the two.[1]

Since our concern here is with what users can do to improve irrigation management, on their own or as part of a more complex system of organization that includes agency staff, agency activities are considered mostly in relation to their support of effective user roles. The analysis in this chapter would apply with appropriate modifications similarly to purely agency-run schemes. The analytical framework for assessing farmer participation possibilities is presented first in summary form. Each of the sets of activities is then reviewed on the basis of what can be learned from the case materials.

The first set of activities focuses on water use:

1. ACQUISITION of water from surface or sub-surface sources, either by creating and operating physical structures like dams, weirs or wells, or by actions to obtain for users some share of an existing supply.

2. ALLOCATION of water by assigning rights to users, thereby determining who shall have access to water.

3. DISTRIBUTION of water brought from the source among users at certain places, in certain amounts, and at certain times.

4. DRAINAGE of water, where this is necessary to remove any excess supply.

These activities apply and must be dealt with at every level of a system, as analyzed in the following chapter. Farmers may be active in any or all of these tasks, directly or through representatives, at any level.

Other activities deal with structures for water control. There is already a standardized classification for delineating such activities with regard to physical structures:

1. DESIGN of structures such as dams or wells to acquire water, channels and gates to distribute it, and drains to remove it.

2. CONSTRUCTION of such structures to be able to acquire, distribute and remove water.

3. OPERATION of these structures to acquire, distribute and remove water according to some determined plan of allocation.

4. MAINTENANCE of these structures in order to have continued and efficient acquisition, distribution and removal of water.

Each of these activities relates to and facilitates the preceding water use activities. They are as relevant to organizational structures as to physical ones. While the structures required for acquisition, distribution and drainage of water are basically physical, those for its allocation are essentially legal or contractual. A capacity for allocation needs to be planned, established, operated and maintained just as surely as does the capacity of a reservoir or a drainage system. Even if allocation activities are not as material as those for acquisition, distribution and drainage, the parallels in terms of the activities involved are substantial, as seen from Table 2.

Table 2: RELATION BETWEEN STRUCTURE-FOCUSED AND WATER-FOCUSED ACTIVITIES

	ACTIVITIES ASSOCIATED WITH WATER CONTROL			
WATER USE ACTIVITIES	Design	Construction	Operation	Maintenance
Acquisition	Design of Acquisition Structures	Construction of Acquisition Structures	Operation of Acquisition Structures	Maintenance of Acquisition Structures
Allocation	Decisions on Water Allocation	Establishment of Water Allocation System	Operation of Water Allocation System	Maintenance of Water Allocation System
Distribution	Design of Conveyance and Control Structures	Construction of Conveyance and Control Structures	Operation of Conveyance and Control Structures	Maintenance of Conveyance and Control Structures
Drainage	Design of Drainage Structures	Construction of Drainage Structures	Operation of Drainage Structures	Maintenance of Drainage Structures

One sees from this how various irrigation activities are undertaken both with reference to a particular phase of water use -- acquisition, allocation, distribution, or drainage -- and to accomplish some kind of control over water in these different phases. The structures involved may be physical, legal or organizational, but all kinds of structures need some design or planning, some construction or implementation, some process of operation, and some activities of maintenance.

Going along with each of these activities which focus on water or control structures are certain organizational activities that marshall human efforts, to make collective action more predictable and effective. These activities can focus on the structures, on the resource of water, or on the irrigation organization itself.

The four basic organizational activities, already introduced in Chapter 1, are:

1. DECISION-MAKING: This applies to acquisition, allocation, distribution or drainage of water; to design, construction, operation or maintenance of structures; or to the organization which deals with these activities. PLANNING is one major form of decision-making.

2. RESOURCE MOBILIZATION AND MANAGEMENT: This involves the marshalling as well as application of funds, manpower, materials, information or any other inputs needed for the above activities, or for any general organizational tasks.

3. COMMUNICATION: This concerns the needs and problems in any of the activity areas noted above, conveying information about decisions made, about resource mobilization, about conflicts to be resolved, etc. to farmers or any other persons involved in irrigation. One purpose of communication may be COORDINATION.

4. CONFLICT MANAGEMENT: This must deal with differences of interest that arise from activities of acquisition, allocation, distribution, drainage, design, construction, operation or maintenance, or from organizational activities generally.[2]

One can have more or less farmer participation in general terms (e.g., how much user participation is there in decision-making or in communication?) or specifically (e.g., how much labor is being contributed in resource mobilization for maintenance? or in the conflict management associated with water distribution, who resolves disputes over water rotations?).

Organizational management activities refer both to physical objects like water or gates and to social relations among people. Resource mobilization deals mostly with material resources but also with non-material things like information and ideas. Even acquiring water through dam or pumping facilities is thoroughly socio-technical because decision-making, resource mobilization, communication, and conflict management are intimately associated with the physical structures and resource flows.

The four organizational management activities closely parallel the preceding set of activities aimed at gaining control over water through physical or social structures, as seen from the following comparison:

STRUCTURE ACTIVITIES	MANAGEMENT ACTIVITIES
DESIGN/PLANNING	DECISION-MAKING
CONSTRUCTION/IMPLEMENTATION	RESOURCE MOBILIZATION
OPERATION	COMMUNICATION/COORDINATION
MAINTENANCE	CONFLICT MANAGEMENT

This similarity does not make them identical, however, because the management activities on the right-hand side apply to each of the activities in the left-hand column:

DESIGN obviously entails making decisions, but it also requires mobilizing information, having communication among the relevant actors, and resolving any conflicts that arise over the design itself.

CONSTRUCTION OR IMPLEMENTATION will involve many decisions about how something will be carried out, substantial resource mobilization, much communication and coordination, and reconciliation of divergent interests and opinions as the work is done.

OPERATION requires decision-making about schedules, work assignments, etc., the mobilization of resources like information, labor and funds, regular communication about schedules, resource contributions, etc., and conflict management to the extent there are any disagreements about operation.

MAINTENANCE likewise calls for decision-making, much mobilization of resources, considerable communication, and also handling of disputes over what is to be maintained, how, and by whom.

From this we see how inter-related all irrigation activities are. They can be viewed from any one of the three perspectives. If one's analysis is water-focused, one still needs to look at the structures and the organized efforts that give control over water. If one focuses on the structures, these have to be assessed in terms of how they affect water flows and what organizational activities they require. Alternatively, if one takes an organizational perspective on irrigation management, these activities have little meaning except as directed toward the physical as well as social relationships subsumed.

The "dimensionality" of irrigation management activities is shown in Figure 1 below. Any activity can be viewed from one of three directions, in terms of a principal focus on water, on structures or on organization, but having at the same time some relation to the other two aspects. In Figure 1, we highlight the conjunction of decision-making activities with respect to the maintenance of structures for water acquisition (e.g. a reservoir bund).[3]

Figure 1: MATRIX OF IRRIGATION MANAGEMENT ACTIVITIES

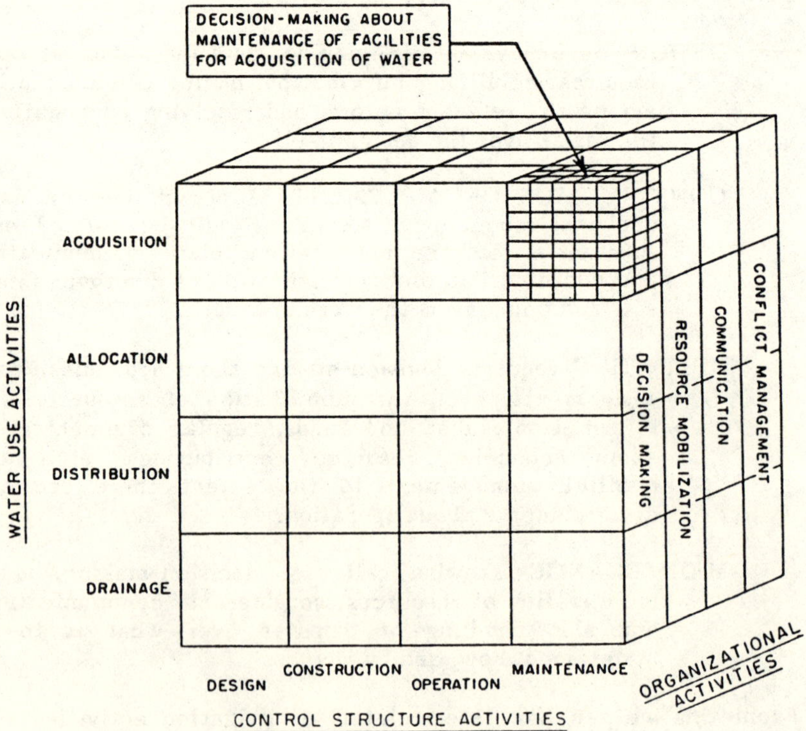

Water Use Activities

Activities focused on water appear to be the most direct forms of irrigation management. This is often seen dramatically in the **acquisition** of water through design, construction, operation and maintenance of facilities such as weirs across rivers, bunds forming catchment reservoirs, or wells tapping underground sources of water. The water thus acquired for use needs to be allocated among uses and among users according to some plan or set of criteria. Water is not assigned to uses or users "naturally," so **allocation** is an essential irrigation activity, even though once a system of rights has been established, there may be nothing to observe. Allocation schemes when they get implemented through the **distribution** of water, on the other hand, are very visible.

A system of conveyance and control structures is needed to transmit and distribute water according to some plan. Like acquisition facilities, canals, channels, regulators, gates, etc. have to be designed, built, operated and maintained. Whether or not **drainage** is an explicit activity in an irrigation system depends on such physical features as topography, soil type and climate. Drainage is frequently neglected as a part of irrigation, but removal of water where this does not occur naturally is crucial for sustained system performance.[4]

What are appropriate or even necessary water use activities for farmers will depend on the nature of the system, and the level at which users as well as agency personnel have responsibility. In Chapter 4, we analyze the levels of operation and organization within a system, and in the subsequent chapter, the various roles which farmers and professionals can perform in irrigation management. Making distinctions of where water use activities occur and who undertakes them will refine the analysis. But consideration of farmer participation in water use activities can proceed here with these other dimensions of analysis having been noted as relevant.

In a simple one-level irrigation system, depending on a run-of-the-river diversion barrage, a small catchment dam or a deep tubewell, and serving perhaps several dozen acres, activities of **acquisition** are usually carried out entirely by the users. In contrast, within huge systems such as those in India, Pakistan, Egypt or Sudan, farmers' involvement in acquisition is quite different. It may involve lobbying with the agency that controls the water to get both allocation and distribution of water to a particular part of the command area. Acquisition may require farmers to protect the water allocated to them to make sure it reaches their channels, as water is sometimes acquired by stealing what has been allocated to others.[5] Acquisition thus can involve quite different activities depending on the size and complexity of the irrigation system and on the level at which the activities occur. When focusing attention on farmer activities in irrigation management "above the outlet," Chambers (1984) finds that organizational strategies for water acquisition are more important than are technical means of acquisition in larger systems. In their analysis of irrigation experience in Thailand, Plusquellec and Wickham (1985:49) observe that the "primary purpose" of farmer organizations in user-managed systems is "to acquire water at the source."

Allocation likewise operates differently depending on the level of the irrigation system where management activities are occurring. At the lowest level, it is a matter of determining how water will be shared among users, according to some rule or criterion. At higher levels in a large system, allocations are more gross, as amounts of water are assigned to areas within the system according to measures like "cusecs per acre." Whether or not farmers participate in determining such allocations depends on the role and authority of agency managers. There may be a system of allocation on paper which is not the system actually governing distribution.

The implementation and operation of an allocation scheme -- that is, **distribution** -- has to contend with the universal problem of locational advantage, where upstream users have greater opportunity to obtain their share (or more than their share) than do users downstream. In the absence of organization (or of powerful users downstream), inequalities in distribution are frequently observed, and organizations are not uniformly successful in promoting equitable allocation and distribution. One of the main purposes of having water user associations is often to deal with this problem.

Because people's livelihoods may be at stake when water is very scarce, distribution becomes more difficult under such circumstances. Indeed, there may be two different schemes of allocation and two different regimes of water distribution between the wet season and dry season, as reported in the Philippines (Robinson, 1982), Indonesia (Duewel, 1982), Mexico (Downing, 1974), and Peru (Mitchell, 1976).

Farmers' ability to handle distribution when water is scarce depends on numerous factors. In many countries, users have devised ingenious means, both technical and social, to distribute water equitably. Examples would be the proportioning weirs found in the hill systems of Nepal (Yoder, 1986) and the subaks of Indonesia (Geertz, 1967), and the traditional practice in some schemes in Sri Lanka and the Philippines of locating at the tail-end of the command area the plot of land which is given as compensation to the person responsible for distributing water (Leach, 1961; Siy, 1982).

There are situations where farmers would like to have someone with authority from outside the community handle the tasks of distribution, as a way of reducing conflicts or avoiding laborious efforts. Lowdermilk et al. (1975) report that Pakistani farmers were glad to have an outside agency distribute water among groups, according to strict rules. However, they preferred to handle water distribution within their groups themselves, modifying the official warabandi rotation by making ad hoc informal arrangements for water-trading (which the agency had declared to be illegal) in order to better meet farmers' individual needs.[6]

Users should be able to distribute water among themselves, though this does not mean they will always handle this task, as there may be competing demands for their labor or social conflicts may interfere. In Taiwan, where farmer responsibility has been taken to the highest organized levels of any country, it is reported that the irrigation bureaucracy now sometimes employs temporary laborers to distribute water at the field channel level on behalf of farmers when they do not want to do this themselves (Moore, 1983). One should not generalize for all levels or all countries or all times about what users can and will do.

We found little reference in the case studies to farmer involvement in **drainage** activities. This probably reflects one of the "blind spots" in the literature, as discussions of irrigation commonly neglect drainage. Improving water control among users can help to alleviate drainage problems so this activity should not be viewed in isolation. Any re-use of drainage water, something farmers are often better able to plan and manage than are engineers, can often increase irrigation efficiency. Certainly where conditions require drainage activities, they must be performed by someone.[7] So an analytical framework should include consideration of drainage, whether or not it presents problems for users and/or agency personnel to deal with.

Control Structure Activities

The **design** of structures to control water from a source (acquisition) to farmers' fields (distribution) and beyond (drainage), according to some guiding structure of rules and procedures (allocation), can be done by users or by various specialists. Generally the larger and the more complicated the system, the more the latter are needed to introduce scientific principles and technical information into the design. However, this does not mean that farmer participation in design activities should be limited to small schemes.

In Chapter 2, we cited some examples of technical prowess on the part of farmers. In the Quinua study from Peru, one finds impressive ways in which irrigation systems have been designed to serve land in a vertical series of environmental zones, each with different conditions. A limited amount of water is distributed in a parsimonious way at successive times to different places at different altitudes to be used for different purposes "in a most economical dovetailing of functions" (Mitchell, 1976:39).[8] The largest system in our sample designed without agency involvement, Chhatis Mauja, was established 150 years ago in the plains of Nepal and covers 7,500 acres. It presented some difficult technical problems because of the rapid flow of the Tinau river during the monsoon season, but the users' design of an organization for operation and maintenance is probably even more impressive than the engineering aspects of their system (P. Pradhan, 1983, 1984).

Construction is a task frequently undertaken by users, often under the direction of technical personnel. The ingenuity of the Marakwet in constructing their schemes in Kenya -- including suspending channels 15 feet above the ground -- was described in Chapter 2. Of course, the technical activity of construction is not possible without the organizational activities involved in resource mobilization that make construction possible. Users often modify the structures in systems designed and constructed entirely by agencies, which constitutes re-design and reconstruction. This may or may not improve system performance.

Operation and maintenance (O&M) are generally treated together, perhaps because they so often occur concurrently, though they are in fact quite separate activities. We will be discussing farmer participation in operation and maintenance when considering irrigation management at different levels in the next chapter, and in relation to specialized roles, considered in Chapter 5. It suffices here to highlight again the fact that maintenance of control structures should be distinguished from the management of the water within the structures. Fleuret observed in his study of irrigation in the Taita Hills of Kenya that, "For the most part, different social groups undertake the two tasks (management of structures and management of water) in different ways at different times" (1985:110).

Organizational Activities

Individuals make decisions and they mobilize and manage resources, but there are limits on what can be accomplished without collective action. By their very nature, communication and conflict management require involvement of more than one person. In irrigation management, the activities of decision-making, resource mobilization and management, communication, and conflict resolution encompass the main focuses of common effort among users. Moreover, they represent also the ways in which users can participate in irrigation management at higher levels within the system, as will be discussed in the concluding section of this chapter.[9]

Decision-making involves more than simply deciding on a course of action. It requires evaluation of performance, identification of problems, gathering information, formulating alternative solutions, building consensus, with decision-making consolidating the various activities. This process can be:

* formal or informal,
* routinized or ad hoc,
* undertaken by all the persons affected or by their representatives,
* binding on all concerned or advisory only.

The decisions taken can vary in terms of their being:

* routine or innovative,
* major or minor (according to whether substantial or small commitments of resources are involved), and
* independent of other decisions or contingent on them.

The kind of decision-making structure that exists does not necessarily correlate with the size and complexity of the system. The Sonjo in Tanzania have a very elaborate "traditional" structure of decision-making for managing their small-scale systems. They have four categories of system membership, but only one category, the major elders, has any decision-making authority. In the "modern" Mwea system in Kenya, an elaborate structure has been established with a Tenants' Liaison Council and Tenants' Advisory Committees, but still there is little user control over water decisions. These are handled through the line of administrative authority which descends from the National Irrigation Board.

The frequency and kind of decision-making needed can vary among systems. Qanat systems found in Iran, Oman and other countries of the Middle East and North Africa are said to run practically "on automatic," with Islamic precepts guiding users' behavior (Spooner, 1974; Wilkinson, 1977; Sutton, 1984). Village authorities become involved with irrigation matters when such systems require repairs, but little O&M activity is needed as the systems are mostly underground, operating according to designed capacities and receiving practically no maintenance. The significant decisions for such systems are the initial ones concerning investment and construction or infrequent decisions about rehabilitation.

The ahar systems in Bihar state of India require more management effort but decision-making is handled informally by local elites. These systems are created by a network of catchment reservoirs connected by distribution-drainage canals (pynes). Prominent farmers whose land is served by reservoirs at the head of the system are expected to mobilize labor periodically from all the areas served to do maintenance work on the pynes. They organize and oversee communal labor whenever it is needed. These examples of minimal decision-making, however, indicate that there is always at least some decision-making activity. The question is, who will do it, and how?

Resource mobilization is the most visible organizational activity in irrigation management, directed most dramatically toward construction as a one-time effort or more commonly to maintenance as an ongoing activity or for rehabilitation. **Labor is the resource most extensively mobilized, though money and materials are also important; likewise farmers' information should be regarded as a major available resource.** One of the best examples of possibilities for resource mobilization is the Nepal system mentioned above, Chhatis Mauja. Its 4,000 farmers contribute 60,000 man-days of work annually for desiltation and maintenance of their main canal. In addition, area and village committees organize operation and maintenance activities within their respective jurisdictions. The smaller systems of Argali and Chherlung in the hills of Nepal mobilize 1,500 to 2,500 days of labor each year (10 to 30 man-days per acre), and both organizations have also raised cash from members to line their canals with cement.[10]

Lesser contributions of resources may be needed in other systems less vulnerable to damage and less beset by siltation. Farmers at San Pedro de Atacama in Chile contributed each month one man-day per hectare for main canal maintenance while also paying a fee of 2 pesos per hectare to cover the ditch tenders' fees. Substantial labor mobilization is similarly reported for traditional schemes in Peru and Mexico (Mitchell, 1976; Downing, 1974).

Even an impressive amount of resource mobilization will not produce the expected benefits unless there is good resource management. This can be provided by users in many circumstances. For example, in the Seraphi system in Thailand, labor responsibilities are assessed quite precisely according to area served, and careful records are kept to ensure that all make their contribution.[11] Very precise and skillful resource management is also documented for the zanjera systems in the northern Philippines. As a spur to performance, for example, there is competition among work groups assigned to rebuild respective sections of the weir, to see which can do the best job. Management tasks may be handled through specialized roles like "irrigation headmen" or through other institutions. Among the El Shabana in the huge Daghara system in

Iraq, contributions of labor and cash for irrigation operation and maintenance are collected through the structure of "tribal" organization rather than through explicit irrigation associations.

Whether it is easier for users to contribute labor than cash will depend on their circumstances. Where money incomes are low, farmers usually prefer providing labor, but when there are good opportunities for wage employment or other claims on farmers' time, they may wish to make payments instead of participating in work parties. Some form of labor mobilization is common in most systems, at least to deal with O&M requirements at lower levels. Still, there are limits to how much of the costs of irrigation management can be covered by labor contributions.

Some interesting "hybrid" systems of resource mobilization can occur. In the Nam Tan system in Laos, farmers paid the "traditional" irrigation headman 16 kilograms of rice for every hectare covered by his services and paid the irrigation agency an additional 80 kilograms of rice for every hectare for its expenses. This amounted to 5 percent of average output.[12] Similarly, the Dharma Tirta irrigation organizations in Indonesia combine contributions of labor and cash, with the fee paid at harvest time and in relation to farmers' yields (Duewel, 1984). Some portion of the funds raised is set aside for further investment, and those organizations which undertake major investments to improve their systems can win prizes from the government for their accomplishment.[13]

There is often resistance reported to requirements of cash payment from users. But Singh (1983) reports that 80-85 percent of farmers in the large Pochampad system in Andhra Pradesh, India were agreeable to a charge provided that the money would be used to improve their service and would not be misused. The O&M requirements of the system were such that some of the resource mobilization needs could be met with labor for maintenance work below the pipe outlet. With 2,000 water user associations (Pipe Committees) in the scheme, Singh estimated that work worth $1.6 million annually could be covered by farmers.

One ingenious method for mobilizing cash directly to meet certain organizational needs is reported in the Izki system in Oman, where water is allocated not according to land area but according to "shares." Although most of these are owned by families, the organization holds some shares which are auctioned on a weekly or annual basis to meet expenses. Capital was mobilized for expansion of the Chherlung system in Nepal by selling off water "shares" in a similar manner.

Mobilizing resources can be facilitated by having a clear and acceptable division of responsibilities between water users and the agency. A good example is the arrangement worked out for pump irrigation in the Senegal river basin, where water users cover the costs of land development, cropping, pump operation and maintenance by contributing labor and money. The government agency (SAED) provides technical assistance and supervision, while a donor agency (USAID) contributed the capital costs of the pumps. Work groups of users appear to be handling their part of the resource mobilization quite satisfactorily, partly because all internal responsibilities of organization and conflict management are left to the group (Fresson, 1979). Without such a clear division of responsibility, farmers and agencies may both leave certain tasks for the other to do.

Resource mobilization from users appears likely to be more successful when decisions on means and shares are left up to the users themselves. Lees (1973) in her study of 24 communities practicing irrigation in the Oaxaca state of Mexico found a great diversity of methods. For example, only nine communities levied direct charges on farmers for water according to their land area or the time water was received; others had more complicated systems which took into account people's contributions of time to village activities, etc. The two systems in Thailand studied by Abha (1979) had a number of different means for mobilizing resources to cover organizational and investment costs:

(1) cash contributions

(2) grain contributions,

(3) exempting certain persons who had organizational responsibilities from their labor obligations,

(4) fines,

(5) selling shares of water, and

(6) selling land, or conferring usufruct rights.

A standardized system would be less able to tap sources that were accessible and plentiful and from which people were most willing to contribute. Flexibility and diversification are important features for any resource mobilization scheme. Furthermore, schemes for resource mobilization are likely to be more sustainable to the extent that there is honest and efficient resource management.[14]

Communication is an organizational activity "so universal that one doesn't see it" (Robert Chambers, personal communication). Its purpose is to help with coordination, which is vital to the discharge of all other irrigation functions. One of the tangible evidences of this function is the creation of specialized roles, discussed in Chapter 5, to handle communication in a number of systems. The Marakwet in Kenya, for example, have set up a system of communication whereby the bad news of need to mobilize labor for a major repair is relayed to all concerned by appointed "blowers," persons who reside at selected locations and are given special horns. The central committee operating the Chhatis Mauja system in Nepal includes two appointed "messengers" who communicate with the 54 member villages about general meetings and dates for maintenance work. They are given a small cash payment plus grain and the use of a bicycle. Similarly in Thailand, in the Seraphi scheme, assistant irrigation headmen serve as "runners" to transmit messages from the headman to farmers and also coordinate irrigation activities in sections of the village assigned to them.

Failures in communication can exact costs in terms of system operation and maintenance. Studies in irrigation systems in India and Pakistan have found that between 70 and 83 percent of farmers did not know the dates when they were expected to do maintenance and repair work or even the dates when water issues would end (Lowdermilk, 1985:6-7). In these cases, even one-way communication -- to farmers -- was inadequate, and two-way communication which would convey farmer needs and capabilities to system managers was still more deficient. In the absence of organization among farmers, communication among them will be limited, reducing their possibilities for cooperation to utilize available water to best total advantage.

Conflict resolution is difficult to judge because where there is much observable "success," it may be because indirect or tacit efforts to avert conflict did not succeed. Where there is little or no strife, rules and procedures may have been devised that handle problems and disagreements so smoothly conflicting interests are adjusted before they lead to disputes or to blows. Or it may mean there was no clash of interests. In certain situations it appears that the need to cooperate for irrigation can overcome propensities for conflict that exist in the community (Wade, 1982).

Some communities and some cultures appear to have a disposition for conflict. This is suggested in several village studies from Pakistan (Merrey, 1982; Bhatty, 1979). In the Daudzai case, village elders are called upon fairly frequently to settle even armed conflicts. In one portion of the Seraphi system in Thailand, conflict became so severe that farmers stopped cooperating and part of that system went out of operation. An aqueduct system of irrigation in one Papua-New Guinea community required the cooperation of two village wards for its maintenance. Periodically, conflict between the two wards became so great that the aqueduct fell into disrepair and the system of agriculture reverted to separate smaller irrigation systems, until cooperation could be resurrected to reconstruct and maintain it for a time.[15]

In certain societies, on the other hand, there appears to be some aversion to conflict. Farmers in Abu Raya, Egypt are reported to want to avoid conflicts within their communities, though this did not rule out conflicts between communities. There was also little conflict within villages in the Oaxaca state of Mexico according to Lees (1973), but disputes were observed between villages, usually over land rather than over water. Similar efforts to maintain good relations among people within a community are said to keep the level of conflict low in Daghara, Iraq, where people in irrigation communities are all from the same tribe. Conflicts which do arise are mediated by community members who claim descent from the Prophet Mohammed. There is strong conflict between tribal groups at Izki in Oman, but all depend so much on the qanat that serves them that they cooperate to keep the system working -- though it is reported that twice, fighting between two Izki groups almost wiped out the qanat.

When there are absolute shortages of water, the likelihood of conflict usually goes up. In the Diaz Ordaz system in Mexico, when water becomes quite scarce during late October, conflicts become quite intense. The system of allocation is accordingly changed between seasons to take into account the different degrees of water stress on the organization (Downing, 1974). The complicated system in Quinua, Peru, described above, was able to operate in the past with very small quantities of water and managed conflicts reasonably well through a traditional hierarchy of "civil-religious" roles. After these roles were abolished by law in 1970, however, the irrigation system became "acephalous" and was prone to much conflict. The legally-recognized "modern" municipal officials have been unable to govern water use, and the strongest individuals and groups are now able to dictate water distribution.

Conflict management may work better through informal mechanisms, such as provided by "traditional" roles and institutions, than through legalistic ones. So long as the authority of the tribal elders among the Marakwet in Kenya and the Sonjo in Tanzania remained intact, conflicts have been managed with little difficulty. The traditional irrigation headman role in Sri Lanka, the vel vidane role documented by Leach (1961), was abolished in 1958 and his responsibilities were vested in elected bodies of water users (Cultivation Committees). These were in turn abolished in 1977 and replaced by an appointed Cultivation Officer (Moore, 1979). Not only conflict management but most other irrigation activities have suffered, with the result that the government is now seeking to establish water user organizations building on old and new "traditions."[16] The key element for effective conflict management roles is not whether they are "traditional" but whether they enjoy the confidence of water users. Imposed roles from outside are not likely to have this, though roles evolved with users' knowledge and cooperation could.

Conflict resolution as an organizational activity resembles drainage as a water use activity. Because it may not always be necessary, it is taken for granted more easily and more often than other activities. Moreover, the preferred situation is where conflicts, like removing excess water, are handled gradually, naturally and imperceptibly.

External Organizational Activities

Organizational activities have been discussed thus far as "internal" to a particular set of water users, who make decisions about what they should do collectively, who mobilize resources from members, communicate and resolve conflicts among themselves. In fact, each of these activities can be undertaken "externally" with reference to other water users or with officials who operate at higher levels. We will not explore this distinction at length here because it requires consideration of levels of organization, the subject of the next chapter. It should be clear that organizations for irrigation management operating at one level, such as the field channel or distributary, may participate in management tasks at other levels.

Farmers managing a distributary canal, for example, may be involved through their representatives in decision-making that allocates water among canals, something which these particular farmers cannot decide on their own. When facing a major repair problem, users may mobilize resources from a government agency -- heavy equipment or subsidized cement purchases, for example. Some of the communication which an organization undertakes will be with persons or agencies outside its own ranks, and a good share of conflict management activity will involve negotiations with other groups.[17] A number of the cases cited above suggested that conflict involving water is often more serious between communities than within them.

The four organizational activities identified in our analysis thus apply to water management relations both among users at a particular level and with users and officials at other levels. To understand better the possibilities and problems of farmer organization for improved irrigation management, we need to consider the matter of "levels," to identify what kinds of participation can usefully occur where within irrigation systems.

NOTES

[1]Hunt and Hunt (1974) analyze their Mexican case in terms of two systems, physical and social. The first subsumes "the relevant physical environment (e.g., the amount of water available plus the artifacts in and on the ground, dams, canals, sluices, etc.)"; the second covers "the social organization connected with the control of the physical system(s)." (1974: 135). We find it better to distinguish three focuses of irrigation management activity as analyzed below.

[2]How these activities compare with irrigation activities discussed by others previously in the literature is analyzed in Annex 2, pages 165-167. This will be of interest to some readers but we do not want to interrupt the progress of our exposition by presenting it here.

[3]This analytical framework creates 64 "junctures" of activity. Not all will be relevant all the time. In his analysis of user-managed irrigation systems in the hills of Nepal, Martin (1986) is concerned with only 24, since design and construction are no longer relevant (only O&M), and there are no problems of drainage given the slopes and soil characteristics. Acquisition, allocation and distribution are essential for successful irrigation, and all four kinds of organizational activities are provided for either by rules or precedents.

[4]In their analysis of irrigation as a physical process, Eggink and Ubels (1984:136-142) identify capture, distribution, application (of water to the soil), and drainage as the four "stages." We have already stated why allocation deserves explicit attention. Application is usually an individual activity, not requiring organization or participation among users, so it is not included here. Eggink and Ubels give useful comments, particularly on drainage.

[5]In the case from Tamil Nadu, India, for example, farmers have specially appointed persons, as discussed in Chapter 5, to patrol the supply channel to ensure that their tank receives its share of the water issued to a string of tanks off the Hanuman River (Meinzen-Dick, 1984). Such water guards may be prepared to use or to mobilize force to stop any encroachment on "their" water. These acquisition methods as well as paying bribes to officials if necessary are reported also in Andhra Pradesh state by Wade (1982 and 1984a). On the system of payments, see Wade (1982a). Patrols are necessary in hill irrigation systems in Nepal to

minimize landslides' interruption of water acquisition (Martin and Yoder, 1983).

[6]A warabandi system allocates and distributes water among users along a channel according to fixed periods of time. If the flow of water is constant, the amount of water thus supplied is proportional to the amount of time, usually set in relation to respective field sizes. See Reidinger (1974) and Vander Velde (1980).

[7]Johnson (1982:159) in discussing drainage problems in the Indus Valley irrigation systems of Pakistan notes that while "elaborate models" have been developed to determine whether public tubewells could successfully lower the water table, "very little effort seems to have been spent identifying how they should be operated and by whom. In particular, minimal attention was paid to issues of organization and human behavior. No effort was made to educate water users about their role in the system, nor, as indicated by the size of the public tubewells, was there any real appreciation of the difficulties of organizing farmers across one or more watercourse areas." Cited in Sims (1986:100-101).

[8]Mitchell (1976) describes how the canal system begins in moist forest regions and ends in low, dry montane thorn steppes. There are several reservoirs in the system. The two major ones are filled during the night and used during the day to irrigate fields near the central town, with some water for the fields also being drawn from the main irrigation canal. The network of minor canals permits separate use of the reservoir and canal water. A simple system of gates has been designed to control the flow since overflowing would destroy the canals. In this case as in others, we see how quite different systems of allocation and distribution are used in the dry and rainy seasons.

[9]A comparison of how these four categories of organizational activity relate to other analytical schemes is considered in Annex 2, pages 165-167.

[10]The Chherlung organization allocates water on the basis of "shares" that have been purchased, either when the system was first built or subsequently as it was expanded. In 1982, it assessed shareholders 250 rupees per share to mobilize funds to make improvements (Martin and Yoder, 1983). In the rainy season, farmers take turns walking daily patrols (in pairs) along the channels to watch for signs of damage so that preventive

maintenance can be quickly undertaken. In Chherlung, all members are assigned to one of seven work groups, each responsible for any maintenance necessary during a particular day of the week. This means that there is always a group ready and obligated to work to keep the system operating during the crucial growing period. If emergency maintenance is needed, all members are required to join work parties or to pay a fine if absent.

[11]In two farmer-managed systems in Thailand, 1200 and 2500 acres in extent, similar resource mobilization occurs. For repairing the weir, farmers must bring a certain number of stakes and certain tools, according to the instructions of the irrigation headman. A careful system of supervision and control operates for labor mobilization. Identification cards are given out at the beginning of each work day and are collected at the end of the day by the headman to check on who has worked a full day as required (Abha, 1979).

[12]Responsibilities for labor contribution were not as precisely assigned but traditionally farmers had contributed labor for maintenance under the direction of the headman, and this appears to have continued.

[13]The village winning the 1981 competition, Glonggong, had invested $270-400 per hectare in upgrading its system, with land levelling, consolidation of holdings, lining of canals and installation of cement turnouts to all fields (to assure precision and equity of distribution). Over 70 percent of the area was able to undertake triple-cropping, due to these improvements. With introduction of shorter-season varieties, annual yields of 7.5 - 9 tons of paddy per hectare were obtained (M. Adams, 1983).

[14]These conclusions match those of a broader comparative study on local institutional development and performance (R. Doan et al., 1984: 5-16; Uphoff, 1986a: Chapter 8).

[15]In the past, such periods of breakdown in cooperation had the beneficial effect of allowing fields to lie fallow and regain their fertility. The government has now built a permanent aqueduct which requires little maintenance, so the wards no longer need to cooperate, and to avoid conflict, they take turns using the land irrigated by the permanent structure.

[16]The vel vidane role was already declining in its effectiveness by the latter 1950s. Leach (1961) reports that the

vel vidane in Pul Eliya was himself engaged in ongoing conflict with some of the more important families in the village. The Cultivation Committees had a checkered but on balance positive record of performance, at least through 1973 after which date they became politically appointed rather than elected (Uphoff and Wanigaratne, 1982). The Cultivation Officers which succeeded them were also political appointees. The introduction of water user groups in Minipe (de Silva, 1981) and in Gal Oya (Uphoff, 1985) has found farmers prepared for collective action following earlier precedents.

[17]How "external" communication can differ from "internal" communication is suggested by observations from Pakistan and Iraq that only large landlords have the resources required to cultivate contacts with the irrigation agency. The costs of such communication include money for transportation, hospitality and even bribes (Lowdermilk, Clyma and Early, 1975; Fernea, 1970).

4
Where Can Participation Occur?
Levels of Operation and Organization in Irrigation Systems

Efforts to assess farmer participation in irrigation management have often focused on differences between "large-scale" and "small-scale" systems. There are more obvious possibilities for user involvement in the latter. But "scale" is an ambiguous and difficult variable to use prescriptively.[1] It is more instructive to think in terms of the structure of an irrigation system using the number of levels of operation and organization to delineate patterns of agency and farmer activity in irrigation management. Considerations of structure can tell us much about the complexity and interdependence of the management tasks within an irrigation system.

"Levels" within an irrigation system are socio-technical in that they can be defined both physically and organizationally. First, there are levels of operation which are established by physical points of water control -- such as a gate or diversion structure -- between the water source and the fields. Below any control structure, one can identify a command area which receives water from that point and which constitutes a level of operation. Such an area may be divided, and subdivided, by subordinate control structures, creating lower levels of operation by dividing the water into different flows. The lowest operational level is established by having a structure (referred to as a "turnout" or "outlet") which serves a channel from which a number of fields receive a flow of water that must be distributed among them (such a channel may be called a field channel or watercourse). At any level of operation there is need to acquire, allocate, distribute and possibly to remove water.

Paralleling this are levels of organization which are created socially when some set of persons at a certain level of operation -- water users and/or agency personnel -- engage in the organizational activities described in Chapter 3: decision-making, resource mobilization, communication, and conflict management. To the extent that these activities are being carried out, whether by water users or by government officials, there exists some organization which corresponds to that level of irrigation system operation. Since the activities may be handled informally rather than formally, there is usually at least some degree of organization at each operational level. But the organization may be quite perfunctory and even unsatisfactory.

Analyzing levels from these twin perspectives of (a) operation and (b) organization directs consideration toward whether they mesh -- and how well. The socio-technical nature of irrigation management is manifested in the interaction of these physical and social sets of activity. When planning or assessing water management at any level, one needs to look at the conjunction -- or disjunction -- of these two sets of activity so as to bring together control over water (operation) and over users (organization). Levels are most clearly defined in terms of physical structures, which form a link between water and users.

The approach taken here encourages more comparable references to levels within irrigation systems. It inverts the standard nomenclature of "primary," "secondary" and "tertiary" levels to gain a more uniform unit for analysis and action. The level at which farmers actually obtain and apply water in their fields is the most basic one, and one which all systems have in common.* The standard terminology which regards the main system as the "primary" level and the users' level as "tertiary" (or even lower) is a top-down view which can just as well be reversed. We begin from below, taking the field channel level as the first level for consideration.

Types and Levels of Irrigation Systems

The simplest type of irrigation system has only one level. All farmers in the system share water from a single source, delivered through a common channel without major bifurcations or other control points at which the flow is sub-divided. Instead a single volume of water is shared among users, more or less equitably. These farmers have a common interest in ensuring and possibly enlarging this flow, though there may be conflicts of interest over its distribution. Management of such a system can be accomplished through a one-level organization, which corresponds

---

* Colleagues at Utah State University refer to this as the "unit command area" in their computer modelling of irrigation systems.

to the single operational level. If the number of farmers is small, as is likely the case, this organization may be quite informal.

The Sonjo in Tanzania provide an example of one-level systems of surface irrigation. These village systems receive water from small springs which are independent of one another. Their management is embedded in the village social organization, with a council of elders overseeing all activities and dealing with any disputes. Other examples of one-level systems include tubewells in Bangladesh (Howes, 1984) and the perimeters along the Senegal River which are irrigated through utilization of lift pump systems (Fresson, 1979; A. Adams, 1977; Patterson, 1984).[2]

One-level systems have the greatest degree of farmer participation and control, and conversely, the least agency involvement. Farmers having constructed the physical system will handle all operation and maintenance tasks. This is due at least partly to government agencies' lack of interest and resources to become involved in many small systems, especially if they are functioning satisfactorily. Agencies such as the SAED in Senegal may provide technical advice or credit, but the manpower requirements for agencies to actively manage such small systems are prohibitive.

A key feature of these one-level systems is the relative independence of their water sources. The amount of water taken from one spring, tubewell, river diversion or small reservoir will have little if any effect on other similar systems. Nevertheless it is important to remember that the wells in an area ultimately share a common water table, and river sources can be diminished. If the supply decreases or the demand increases markedly, the water taken by one system may reduce the water available to other similar systems, thereby offsetting the above-mentioned independence.

Two-level systems are found with larger water sources and command areas or where several small systems are interdependent. In such circumstances, a second level of organization becomes necessary to allocate water among the lower-level units and to arbitrate disputes among them. The ahar systems in Bihar state of India draw water from common feeder canals (pynes) and thus need at least informal communication and decision-making. As reported in Chapter 3, they have informal but adequate methods of resource mobilization for maintenance. Such systems may function mostly as one-level operations but in some respects they are two-level systems. They operate through horizontal rather than vertical linkages, i.e. through cooperation among similar systems that have no superordinate decision-making body.

The furrows constructed by the Marakwet in Kenya are intended to supply water to one-level systems. But clans more distant from the water source find that the labor required for digging a channel is so great that several often join together to construct one large furrow with several major diversions to each clan's territory, making this a loose two-level system. The formation of a federation among formerly independent zanjeras in the Philippines, each taking water independently from a river source also provides an example of this process, though if their constituent work groups (discussed below) are counted, their federation constitutes a third level of organization.

Any system having more than one level will be hydrologically and organizationally more complex, with several levels of organization corresponding to operational areas established by physical control points. Within such systems, there will be multiple small command areas and small groups of users, each served by a common turnout structure and field channel. Such a unit command area represents the basic level of operation and of actual or potential organization. These structures establish socio-technical units that resemble one-level systems in many ways but which lack independence of action.

The size of these basic units of operation and organization is determined by considerations of hydrology (the size of the smallest area separately commanded) and by the average field size, as well as by social considerations such as residential proximity, ease of communication, and social homogeneity. The larger the command area and the smaller the holdings, the greater the number of water users in a turnout group. Conversely, having smaller areas and larger holdings will reduce the number of users in the lowest level of organization.

The number of farmers cultivating within what are often called "turnout areas" appears to range most often between 10 and 15, according to a study of small-scale irrigation in the Philippines, Malaysia and Laos (Coward, 1977). New water user groups in a large-scale system in Sri Lanka exhibited a similar range.[3] A study of irrigation in Niger found the number to be 8 to 13 (Laucoin, 1971). The official size of the basic small group in Taiwan Irrigation Associations (where average farm size is very small) is about 50 members, but informally these are further subdivided (Moore, 1983).

Like water users in a small one-level system, all the farmers within a unit command area have a common interest in its water supply and operation, even if there may be competition over the water available within that area when supply is limited. Maintenance tasks are important at this level for facilitating both water access and water control. In practically all systems, large and small, having many levels or just one, operation and maintenance at the base level are the responsibility of water users.[4]

What distinguishes these basic groups in complex systems from water users in a one-level system is that there are a number of similar base groups all sharing water from a common source at a higher point in the system. This usually elicits some degree of organization at that higher level, though it also provides occasion for conflict between base-level groups.

An interesting example of a two-level irrigation system with intermittent organization at a third level is reported in the Papua-New Guinea village of Wamira. Our analytical framework clarifies the problems and dynamics of such a situation. The social system of Wamira -- a village divided into two wards which each in turn encompassed multiple hamlets -- would appear ideal for a three-tier organization for irrigation (village-ward-hamlet). But the wards had a history of hostility toward each other, and each usually operated its own irrigation system, obtaining water from separate rivers. Within each system, hamlet groups had their own off-takes and distribution channels. Since the wards had independent sources of water, there was no need for village-level organization, and two two-level systems operated side by side.

As reported in the previous chapter, the two wards from time to time cooperated in constructing a common aqueduct which supplied a field area they could cultivate in addition to their gardens. For this they would establish an informal organization at the third level for some time, involving extensive resource mobilization in the construction phase, through ad hoc decision-making and communication. However, the inability of the organization at this third level to manage conflicts between the wards meant that maintenance would eventually stop and the aqueduct would fall into disrepair.

The field area in Wamira would then then go back into fallow until some new but short-lived organization at the village level reconstructed the channel. This cycle was ended when the government built a permanent aqueduct that does not need any maintenance by users. No third level of organization has existed since then in Wamira, even intermittently. The two wards have agreed between themselves simply to alternate use of the field area, so they do not have to cooperate at all on irrigation management. The system can operate without any further decision-making, resource mobilization, communication, or conflict management at the third level of organization.

Three-level organizations for irrigation represent additional complexity and possibilities. The Sananeri tank irrigation system in Tamil Nadu state of India is an interesting example, having informal organization at the lowest and highest levels. Most of the decision-making, resource mobilization, communication and conflict management are focussed on the tank (which is filled by a channel from the Hanuman River) and on its command area of 440 acres. At this second level of operation and organization, one finds officers and a treasury plus staff hired by the tank association.

The command area is divided into three sections, each constituting the first level of operation, served by its own channel which is fed by one of the tank's three sluices. The tank association employs for each section two ditch-tenders (niirpaaycci) who apply water to fields according to an allocation scheme agreed on at the tank level, so there is no need for any formal organization at this first operational level. But there are patterns of informal decision-making, resource mobilization, communication and conflict management, and organization of a rudimentary sort can be said to exist.[5]

In an upward direction, the Sananeri association cooperates with user associations for the other tanks served by the same canal from the Hanuman River, establishing an informal third level of organization. The associations coordinate their off-takes of water to fill their respective tanks from issues given by the Public Works Department (PWD), which manages the main river system. (The PWD-managed river system constitutes a fourth level of operation and organization, as discussed below.) From time to time, delegations of representatives from all the tank associations will approach the PWD for special water issues or for needed repairs to the canal. Although the PWD is responsible for maintaining the canal, groups of farmers sometimes do maintenance work on it, using their own funds and labor, partly to ensure regular supply and partly to build up good will with the PWD to enhance their bargaining position on water allocation (Meinzen-Dick, 1984).

The formal three-level organization managing the Chhatis Mauja scheme in Nepal, discussed previously, actually has an informal higher level, similar to the third level of the Sananeri system just described. There are 54 village committees at the base, joined in nine area committees, which in turn make up a central committee for the 7,500 acre area, with officers directly elected by the cultivators. Because three other user-managed systems also draw off water from the Tinau River, the systems have established some joint communication, decision-making and conflict resolution. But because they have independent off-takes, there is no common resource mobilization at this fourth level.[6]

Beyond three levels of organization, the role of users becomes relatively attenuated in any large system. The Pekalen Sampaen system in Indonesia, for example, covering almost 700,000 acres is made up of 139 sub-systems. These average 5,000 acres each, but range from 70 to 43,000 acres, having between one and four levels themselves. The whole system is divided into three large districts, and further into eleven sections, and then into sub-sections, and sub-sub-sections. Water users are actively involved in various management activities from below at the first and second levels, and sometimes at the third, but not higher.

The Lower Lalo system in the Philippines, irrigating 7,000 acres, is an example of a relatively small scheme with more than three levels of operation and organization. For management purposes, the system is divided into five "districts," which are subdivided into "zones" -- ten in all. At the lowest level of operation and organization are the "rotation areas" of which there are 93 in the system, averaging 75 acres each, a convenient and manageable base unit.

The Muda scheme in Malaysia is an example in between. It has 200,000 irrigated acres, of which 85,000 acres were covered by farmer organizations by the middle 1970s. The Muda authorities adapted many features of the Taiwan Farmers Association model, and thus set up small "work groups" of 7 to 10 farmers who cultivated about 50 acres as the basic unit. These groups are aggregated into "Small Agricultural Units" (as in Taiwan) which encompass a single village or several villages with about 150 acres. The SAUs in turn are each attached to one of 27 Farmers Associations, a third level of organization set up by the Muda Agricultural Development Authority, the agency in charge of the system. It appoints general managers and staff for the associations, each of which has a representative on a Board of Directors (fourth level) for the entire scheme (Afifuddin, 1978). The FAs average about 3,200 acres in area, an appropriate size for third level organization.

In parts of the Indus river valley irrigation network in Pakistan, the size of first-level command areas are on average 400 acres, several times the "norm" discussed below. Size of holdings is rather larger in this system than most of the other irrigation schemes found in other developing countries, and the intensity of cultivation is considerably less, the systems having been designed more for drought protection than maximizing yields. We note that there have been suggestions for reducing the size of the watercourse area, by redesigning the watercourses and increasing their number (Merrey, 1983).

The portion of the 250,000 acre Daghara irrigation system in Iraq studied by Fernea (1970) is interesting because its local organization for irrigation is part of the more comprehensive social organization of the El Shabana "tribe." Water acquisition is not a problem for users in the area since the government built the system to provide water to everyone. There is little decision-making to be done by water users. At the lowest level, small groups called <u>fakhds</u> with about 10 to 30 adult males make all decisions by consensus, with no specialized roles for water management. Resource mobilization, communication, and conflict management all occur regularly but informally at this level. At the next higher level, larger groups (shabba) with 20 to 50 adult males cooperate in water allocation, maintenance, conflict management, etc., still with no formal leaders for irrigation (though sada, descendents of the Prophet, are supposed to encourage productive and amicable relations among users by setting a good example). The third level of tribal organization, the ashira, is much larger, with 500 to 900 adult male members. It has a socially and legally recognized leader (shaykh) who oversees irrigation activities and also represents the community in dealings with the government irrigation bureaucracy. Beyond this level, water users are involved only indirectly or informally in water management. This is the pattern we find generally with large irrigation systems. Figure 2 below depicts a number of systems which range from one to four levels.

Figure 2: SCHEMATIC COMPARISON OF IRRIGATION SYSTEM STRUCTURES, BY NUMBER OF LEVELS OF OPERATION AND POSSIBLE ORGANIZATION

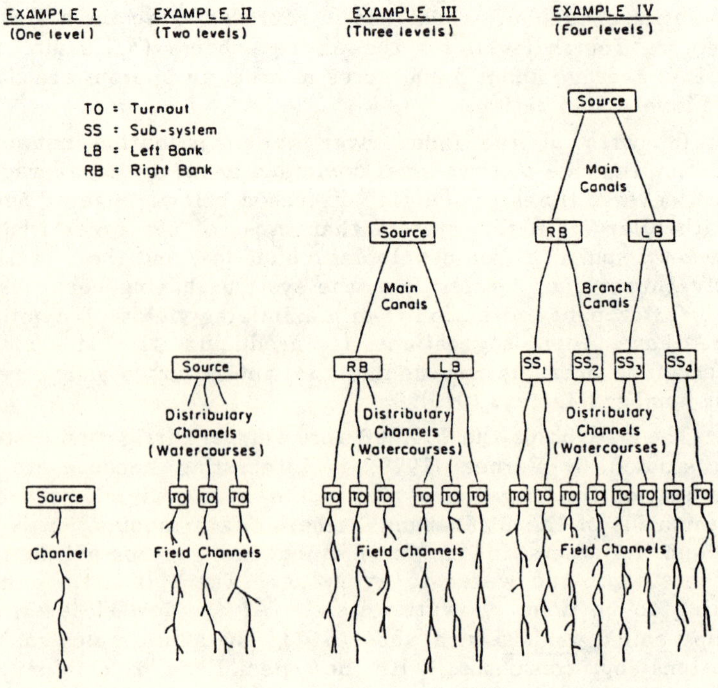

Rationale for Analyzing Levels of Operation and Organization

This way of looking at levels inverts the way in which irrigation systems are usually described now, in purely technical terms and from the top downwards. The highest level, corresponding to the area served by the total supply of water is called the "primary" level by engineers and planners, with subdivisions in descending order referred to as "secondary," "tertiary," and even "quaternary." Depending on the size of a system, of course, these levels can represent quite different hydrological and sociological realities. In Pakistan, the "tertiary" level of community-managed irrigation systems in the Northwest Frontier Province is small and compact, with complete farmer responsibility for management. It bears little resemblance to the "tertiary" level of systems in the Indus Valley of the Punjab, which will be a hundred times larger and entirely under agency control.

The analysis here, by starting with delineation of "first-level" hydrological and sociological units of operation and organization at the base (unit command areas), no matter what the overall size or structure of the irrigation system, offers a more comparable way of speaking about "levels." Where a "second level" as defined here exists, its functions will be more similar from system to system than are the "secondary" levels of irrigation systems which as conventionally classified can encompass areas anywhere between 100 and 100,000 acres.

The nature of irrigation management tasks will vary between lower and higher levels of irrigation systems. Although no definitive rules hold across all systems, some general tendencies can be noted, such as the generalization that farmers are most active and have most responsibility at the lower levels, while agency responsibility will be greater as one comes to "higher" levels.

There is no necessary point at which responsibility gets taken over by the agency. In the Chhatis Mauja system, the third level organization interacts with the government but farmers carry out all management functions for the 7,500 acres. Indeed, when officials tried to intervene in system management, the farmers' organization challenged it to assume full responsibility for the system or to desist from what users saw as "interference." The government chose the latter course. At the other extreme is the Mwea system in Kenya where the system was planned for virtually no user participation in irrigation management even at the lowest level; water was to be delivered to farmers' fields by agency personnel. This has proven to be not only exceedingly costly but also unconducive to good irrigation results.

One observable tendency is for higher levels of organization to operate more formally, although informal procedures and consensual norms will exist even at the highest levels. When the number of persons involved in irrigation management is larger and the distance between points in the command area is greater, more explicit decision-making processes and written records become necessary. Social pressures which could elicit cooperation in small groups with more informal modes of operation are no longer as effective. When more people are involved and there is less frequent and less personalized interaction, explicit procedures and formalized sanctions will usually be needed to sustain widespread coordination of activities.

Direct participation of farmers in collective decision-making is more feasible at lower levels. The area and number of farmers encompassed by a level increases as one goes "upward" in a system, from first to second or from second to third levels. As this happens, engaging all water users in deliberations becomes more difficult to arrange. Instead, representatives of the lower levels are likely to function as part of the higher-level organizations, in a form of indirect participation. Executive or committee styles of decision-making become more common than assemblies of all members.

Land and labor contributions as a mode of resource mobilization for irrigation management are more feasible and useful at lower levels; they are usually more difficult to mobilize and manage at higher levels of operation. The Chhatis Mauja case shows that with good organization, "in-kind" resources can be mobilized and used effectively even in a three-level system, so it is possible to accomplish a great deal with non-monetary resource mobilization. Still, at higher levels, cash becomes more important, with salaries and purchases of materials and equipment necessitating some monetary contributions for organization and operation.

The nature of irrigation management tasks will change according to the level within the system where they are addressed. In large schemes, for example, what is regarded at the lower levels as water "acquisition" corresponds to higher-level activities of water "allocation and distribution." Also in larger systems, water users will have to give relatively more attention to the tasks of operating and maintaining their organizations -- compared to managing physical structures -- because organizational arrangements will have to be more complex. Further, in larger systems, users' organizational management activities will be directed relatively more toward what goes on at other levels, and particularly toward agency personnel involved in irrigation management.

The correspondence between technical levels of operation and social levels of organization in a system should always be considered. One recurring issue is whether it is better to join together residential neighbors or "field neighbors" -- cultivators who share a common source of water rather than being members of a social community (Coward, 1980). We consider this question in Chapter 8, but note here that the set of persons in the organization should usually be those who are involved in the operation at that level, whether or not they happen also to be residential neighbors.

Where holdings are fragmented and held in several areas of a system, it may make sense for work groups to be formed without respect to where the farmers' landholdings are located, as reported in some systems in the Philippines (Siy, 1982), India (Sengupta, 1984) and Mexico (Downing, 1974). Another reason for not organizing groups always according to members' proximity to one another as cultivators is put forward in the Muda scheme in Malaysia. There it was hoped that mixing upstream and downstream farmers in the lowest-level work groups would foster a spirit of cooperation instead of competition between head and tail. Whether this had the intended effect is not reported (Afifuddin, 1978).

Where new irrigation systems are being constructed in an area that is already settled and cultivated, attention should be given to designing the system to serve existing social groups. This was done to some extent in the construction of certain large canal systems in India which provided an outlet for each village (Wade, 1979). Studies in Pakistan show the advisability of this by a negative example, where watercourses in the Punjab were laid out without regard to the pattern of landholdings among kin-based "brotherhoods" (biradaris) which formed the basis of social organization. The potential for cooperation to emerge from existing patterns of decision-making, resource mobilization, communication and conflict management was not capitalized on, and it has been difficult to obtain cooperation between members of different biradaris who cultivate along a given water course, even when all would benefit from working together.[7]

Relation Between Number of Levels and System Size

Recognizing that "levels" in irrigation systems can be understood both in organizational terms and in operational terms is important, and achieving a correspondence between these social and technical sets of activities is crucial for satisfactory system performance. From our analysis of cases in the literature, we have observed that differences in the number of levels found in irrigation systems can be roughly characterized in terms of orders of magnitude with regard to the size of command areas. While there is variation due to natural factors like topography as well as due to man-made causes like engineering design and land holding patterns, generally speaking, we find the following:

(i)	One-level systems or the first level in larger schemes (unit command area)	Generally under 100 acres--e.g. the Matam pump schemes in Senegal, ranging from 40 to 65 acres, or the phads in large schemes in Maharashtra state of India, from 20 to 100 acres.[8]
(ii)	Two-level systems or the second level in larger systems	Generally over 100 and up to 1,000 acres--e.g. the hill irrigation systems in Nepal, or the Sananeri tank system (440 acres) in Tamil Nadu state of India. Rotational areas in Taiwan Irrigation Associations are 125-350 acres, while the first level "teams" below them cultivate 40-50 acres.
(iii)	Three-level systems or the third level in larger systems	Generally over 1,000 and up to 10,000 acres--e.g. the Chhatis Mauja system (7,500 acres) in the plains of Nepal, or the Farmers Association level (3,200 acres) in the Muda scheme in Malaysia. The third level ("sector") in the Mwea irrigation system in Kenya covers 2,000-3,000 acres, for example.
(iv)	Four-level systems or the fourth level in larger systems	Generally over 10,000 and up to 100,000 acres--e.g. the scheme managed by the S.Y. Farm Land Improvement Association (23,000 acres) in South Korea, or the Left Bank system (62,000 acres) in the Gal Oya scheme of Sri Lanka.

Beyond this, the relationships become looser. The whole Gal Oya scheme (120,000 acres) has five levels of operation and should have five levels of organization when the current program of introducing farmer participation is completed. The Gezira system (2 million acres) in Sudan likewise fits the scheme of analysis in terms of levels of operation (more than five), however it has no corresponding structure of levels of organization. Neither does the Rahad scheme (300,000 acres) also in Sudan, which has fewer levels of operation (four) than expected according to this "orders of magnitude" analysis.

Our intention here is not to propose some fixed or necessary relationships. Rather it is to report a quantitative association we have observed in the number of operational and organization levels in irrigation systems according to size. The mid-point in these ranges is in many ways more descriptive than the range itself. This would make the first level of operation and/or organization typically about 50 acres, while the next level (or size) would commonly be several hundred acres. The third level would be several thousand acres, the fourth level, several tens of thousands of acres, and so forth.

It might be thought that these levels should be denominated in terms of the number of farmers rather than the number of acres. Unfortunately, the number of farmers operating and organized at a particular level will always vary because average holding sizes differ so much across systems. If we used the number of farmers to delimit ranges, we would have to state each end of the range itself as a range, which would make comparisons even more complicated.

In situations where the average landholding size is small, the area subsumed under any particular level will tend to be at the lower end of the ranges described above. This is because the number of user-members would otherwise be greater and also because the larger number of smaller holdings makes operational activities more complicated. Where holdings are large, the converse is true. Fewer users need to be involved and the tasks of management are simpler, so the area can more easily be in the upper end of the range.

While resource mobilization may be easier with larger numbers of water users, the other three organizational activities, decision-making, communication and conflict management, are not. At all levels in a system, some persons -- whether water users or agency staff -- will have to be carrying out, formally or informally, regularly or at least intermittently, the various organizational functions which aim at ensuring that the operational functions of water acquisition, allocation, distribution and drainage are managed productively.

Irrigation Groups

Efforts to improve irrigation management often focus on farmers in their organizational and operational activities at the lowest level, along the channel serving their fields. In some assessments of water management, water waste is blamed on farmers who are themselves the "victims" of poor management at higher levels.[9] Even if one regards farmers as more sinned against than sinning, however, there is usually considerable scope for improvement at the lowest level. A necessary but not sufficient approach to improving irrigation management involves establishing or strengthening base-level organization -- groups of farmers whose fields are served by a common source. Such "turnout groups" provide the "building blocks" for irrigation management structures in any system having multiple levels. Just as physical control structures like gates and pipes are needed to prevent water from just flowing freely at the lower levels, there need to be some social control structures such as groups provide.

The zanjeras in the Northern Ilocos area of the Philippines and the subaks on the island of Bali in Indonesia are well-known indigenous organizations for irrigation management, with impressive capacities for resource mobilization and management as documented by Lewis (1971), Siy (1982), Geertz (1967), and Birkelbach (1973). It is important to note that these are two-level organizations. The members of a zanjera number between 14 and 140, with 40 as an average, cultivating 50 to 400 acres (the average is 140). The effectiveness of the zanjeras depends greatly on the strength of their constituent work groups (gunglos). These each have about 10 members (5 to 25) cultivating some 35 acres (20 to 125). The structure of the subaks is similar, as each subak is made up of tempak groups. Each of these has its own headmen, similar in duties but subordinate to the headman of the subak.[10] These are about the same size as the work groups already mentioned at the base level of the Muda organizational scheme in Malaysia.

The Pipe Committees set up in the Pochampad irrigation system under the Command Area Development Authority of Andhra Pradesh in India were each responsible for an outlet command area (chak) of about 100 acres. Initially, these committees were informal and were made up of 5-6 active farmers who were selected by the irrigation staff and who each represented a 15-20 acre "zone" within the chak. Singh (1984) reports that the committees functioned better when their members were chosen by field neighbors within each "zone." This established better communication upwards and downwards and more sense of mutual responsibility. Once again, the two-tiered structure proved to be important, even though the zones had only informal organizations.

This pattern of base-level groups is widely reported in the literature. In the Chhatis Mauja system of Nepal, the average size of village sub-systems is about 75 members. Within this level of organization there are informal groups, so the system actually has more than its formal three tiers. Within the Irrigation Associations (IA) in Taiwan, Irrigation Groups cover an area of about 375 acres. Each is led by a formally-elected Chief and is supervised by an official from the IA office. As these are too large to handle all O & M duties, they are each broken down into three Irrigation Teams, with informally selected heads. Maintenance budgets have been officially devolved to the Irrigation Groups, but the responsibilities of the smaller Teams are not very clear.[11]

Somewhat larger base-level user groups are reported in Daudzai, Pakistan, but they are broken down into maintenance groups averaging 15-20 members. In the Punjab, groups go as high as 150, though they too often operate with smaller sub-groups. Coward (1979) describes base groups in the Nam Tan project in Laos made up of 40 to 45 field neighbors (range 30 to 60), while in the Mexican irrigation system documented by Downing (1974), most management activities are carried out by groups known as tramos which average about 30 members (range 13 to 41). Leadership of these groups is considered an obligation and all members must serve in the offices of president, scribe and treasurer in rotation. There is no formal organization above the tramos level, though some informal cooperation among groups occurs in construction and maintenance. In the Dharma Tirta organizations being introduced in Indonesia, base groups averaging about 40 farmers are handling a full range of management functions. It is reported, however, that the kelompok basic management area may be subdivided into sub-kelompoks where water is more scarce in order to have smaller management units that can attain better water control (Duewel, 1982).

Some of the most impressive group activity is reported in the Matam area of Senegal, where the agency (SAED) works with newly formed groups that receive pump-lift irrigation for small areas along the river's perimeter.[12] Each group chooses its own chairman and organizes its work as it pleases, handling the day-to-day operations and keeping the pump engine fueled. Land development work is completed in as little as four months, which is judged quite an accomplishment. In the Bakel area downstream, one group grew from 40 to 270 members, leading to the formation of many new groups. In Niger, a three-tiered scheme is based on organizations known as "blocks" with 8 to 13 members (18-25 acres). These are joined into larger "sectors" which in turn are combined into still larger "zones."

The widespread existence and significance of base-level groups in irrigation management underscores what N.G.R. de Silva, now chairman of the Mahaweli Engineering and Construction Authority in Sri Lanka, has referred to as "the power of small group processes" (1984:5). While a Deputy Director of Irrigation, he himself introduced a four-tiered system of farmer organization in the Minipe scheme (15,000 acres) which showed considerable improvement in water management within several years. Though initiative came from above, efforts to increase water use efficiency started with farmer organization at lower levels of the system.

Higher Levels of Organization

The organization developed at Minipe was more than just a collection of field channel and distributary channel organizations. Like most of the more successful cases of irrigation management, the effectiveness of small groups at the lowest level depended not just on the solidarity of members but derived also from their vertical linkage to some higher level of organization, which could similarly establish horizontal linkages and coordination among farmer organizations at the same level of operation.[13]

The creation of a federation combining nine zanjeras in the Philippines added greatly to the performance of the constituent units as shown by Siy (1982). One impetus for their cooperation was a change in the channel of the river from which the zanjeras diverted their water supply. As it was increasingly difficult to obtain enough water, they benefited from coordinating their construction of weirs. The small groups in the Bakel region of Senegal, referred to above, even though their sources of water were not interdependent, set up a federation in 1976 after several years of operating separately. They felt they could deal better with the agency with which they were working (SAED) if they could take some decisions together. Also, joint mobilization of funds and labor was possible, and the groups could help each other keep their respective pumps supplied with fuel and spare parts.[14]

The rich range and variety of irrigation experience reported in this chapter gives substance to the abstract concept of "levels" with which we began. What kinds of farmer participation are possible and desirable will depend on the level of operation within a system. The primary group of irrigators at field channel level (the unit command area) represents the "lowest common denominator" for irrigation management. Users are almost always responsible at this lowest level for the full range of water, control structure and organizational activities analyzed in Chapter 3.

As decision-making, resource mobilization, communication and conflict management occur at higher levels of operation concerning larger command areas, themselves made up of sub-areas operating somewhat independently, the role for farmers will change, and the comparative advantage of agency personnel as irrigation managers becomes greater as a rule. Still, there can be farmer involvement in decision-making, communication and conflict management at rather high levels of systems through representatives (resource mobilization can still be direct). Decisions establishing policies or priorities will be more amenable to user inputs than are detailed technical decisions about water issues and schedules. In very large systems, farmer participation at the highest levels can become so indirect as to attenuate the advantages it offers -- intimate knowledge of local conditions, a strong sense of personal commitment to achieving good performance, and social solidarity and sanctions to support collective action.

This observation confirms the importance of identifying and assessing differences in levels when trying to evaluate or provide for farmer participation in irrigation management. We turn now to a consideration of experience with farmer responsibility for different management activities at various levels in systems described in the literature.

NOTES

[1]The concept of "scale" presents the following difficulties when one tries to use it for analysis and prescription:

(a) Scale measured in terms of command area is not likely to be the same when calculated in terms of the number of farmers in the system. This latter statistic is more relevant to considerations of participation, but the former is the more commonly used standard for classification and comparison.

(b) Determining a standard cut-off point between large and small scale schemes is practically impossible -- as difficult for number of farmers as for command area -- and introducing an intermediate category of "medium-scale" does not eliminate the problem.

(c) Where countries have set arbitrary criteria for distinguishing scale, they are quite different. In most Indian states, "small" schemes are those below 5,000 acres, whereas in Sri Lanka the dividing line is 200 acres, for example. Scale may otherwise be determined arbitrarily by size of investment.

(d) The actual area cultivated may be larger (even much larger) than the officially reported command area -- or it may be considerably less. So the data base for determining "scale" is not as solid as it might appear.

[2]As noted later, these small systems can be joined or federated organizationally even though they are operationally independent.

[3]Of the first 70 groups formed by farmers at field channel level in the Gal Oya project, two-thirds were in the 10-15 range, and all but three were between 8 and 20 members. Where field channels were longer than average, with more than 20 farmers, usually two groups were formed, and if there were more than 40 farmers, three groups. Decisions to subdivide channels for organizational purposes, or to amalgamate small ones, were colored by factors such as whether farm neighbors lived near one another or not.

[4]The one exception to this in the cases we examined was the Mwea scheme in Kenya. In this agency-managed system, farmers were only "tenants" and all irrigation management activities, including delivery of water to each field, were reportedly done by agency employees in the early 1970s. This has been changed subsequently (Chambers and Moris, 1973).

[5]The accountability of the niirpaaycci to farmers is reinforced by the practice of the latter paying the ditch-tenders a set amount of rice per acre served at the end of each season. This face-to-face transaction encourages the irrigation specialists to provide good service because farmers can balk at payment if they have grounds for dissatisfaction. Wade (1979) reports the same practice, for the same reason, in irrigation systems in Andhra Pradesh state of India.

[6]There might be joint claim-making on government by the several associations in the future, to get financial or technical assistance, which would represent externally-directed resource mobilization. But thus far, the Chhatis Mauja organization has

resisted any government involvement in the operation of its system. If one counts the informal user organization below the village level, Chhatis Mauja is a five-level system.

[7]Merrey (1982) shows how a certain cultural predisposition in the community, described as a fierce regard for "honor" (izzat), makes cooperation difficult in any case. Cooperation becomes more problematic and conflict-laden however when it must cross existing lines of social organization.

[8]Patil and Kulkarni (1982:4) report this range, saying that it depends on topography, but adding also: "In exceptional cases, phads of the size below 6 acres are also to be found." We noted on page 65 that the lowest level in some Pakistani systems is as much as 400-500 acres, but this is thought to be too large.

[9]Chambers (1981) cites a statement from the World Bank's 1978 World Development Report: "Wasteful water management and poor maintenance can be blamed in large part on the hierarchy of social relationships among farmers." (para. 40) No comparable criticism was made of Irrigation Departments.

[10]There are also specialized "groups" called pekaseh subak which handle the actual tasks of water distribution and irrigation system upkeep. They are exempted from payment of a cash tax (per tenah of land cultivated) into the subak treasury, and receive a portion of the tax left after general operating expenses of the organization have been met. In the two subaks studied by Geertz (1967), about one-third of the members were in these specialized groups which did daily distribution and maintenance tasks on a rotating basis, and which every two weeks were mobilized to do larger jobs.

[11]Moore (1983) suggests that the Teams do not have as active or effective a role as advertised in working out and implementing crop rotation systems, and he describes the Teams as "shadowy." Moore expresses relief that the systems "were not in fact managed by rigid adherence to the very elaborate, detailed and very time-consuming procedures and institutional arrangements which are implied in 'official' accounts (of system operation)."

[12]A World Bank report says that these groups (of 12-15 household heads each) have been able to do a better job than SAED of maintaining the equipment needed for irrigated agricultural production (Cernea, 1984:11).

[13]A partial exception would be the Mexican case reported by Downing (1974), mentioned above. But even there, the activities of the tramos were orchestrated informally by the sindico, an irrigation official working for the government. Also, because farmers often had land in several locations, they could be members of more than one tramo, and this made for informal horizontal linkage and coordination. In other irrigation systems in the state of Oaxaca, documented by Lees (1973), there was little cooperation between systems, as only two of the 26 villages she studied worked together in maintaining canals. Such limited cooperation among water users seems to be an exception rather than the rule.

[14]Pump technology may seem to create an independent source of water supply, but there is crucial dependence on man-made if not natural inputs which can give impetus for cooperation.

5
Who Participates?
User Roles in Irrigation Management

A wide variety of arrangements can be found for managing irrigation systems, large and small. Since our interest is in farmer involvement, we are looking at the ways in which user participation has occurred and could be productively expanded. The roles and performance of agency personnel concern us as they affect farmers' willingness and ability to undertake management responsibilities.

Participation in irrigation management can be differentiated by kinds of activity and by levels, as analyzed in Chapters 3 and 4. Distinctions need also to be made in terms of who is engaged in the various tasks of managing water, structures and organization, as one should not assume any homogeneous "flow" of participation. One may be as concerned with who is not participating in various management activities as with who is.

Our purpose in focusing on different kinds and degrees of participation is not to propose prescriptive norms. Whose participation is desirable and possible will vary according to the context, discussed in Chapter 6, as well as according to the nature of the activity and the level at which it is occurring. The following characteristics of participants (actual or potential) are identified to alert observers to possible imbalances or gaps in user participation. What might be done to remedy any shortcomings will depend on socio-economic factors, technical constraints, policy objectives, administrative capacity, and so forth.

The following variables should be considered:

(1) **Location:** The most widespread differential in participation in irrigation activities (and even more so in irrigation benefits) is between upstream and downstream water users. One of the aims of farmer organization may be to reduce these differentials by

involving downstream farmers in decision-making, for example, or by establishing forums to resolve competing water claims in an equitable manner.[1]

(2) **Gender:** This is often an "invisible" differential because it is frequently taken for granted that water management is "man's work," even though women are heavily involved in the actual work in the fields. Changing gender divisions of labor is slow and difficult, but biases in participation opportunities should be noted, and where more equitable and efficient outcomes could be promoted by more active participation of women in irrigation management, this is to be supported. (See page 89 below.)

(3) **Landholding:** This has two aspects, differences in the amount or quality of land, and in cultivators' tenure status. For example:

(a) The mobilization of labor resources for construction and maintenance may come from smaller farmers while decisions about water allocation and distribution rest with larger ones.

(b) Membership on a council which resolves water disputes may be restricted to landowners, so that tenants are at a disadvantage in any conflict resolution.[2]

Rates of membership or meeting attendance may be different between large and small operators, or between owners and share-croppers, which affects the outcomes of irrigation management.

(4) **Economic Activity:** This also has two aspects to be considered:

(a) Where a substantial number of cultivators are only part-time farmers, and are engaged in other activities, their stake in operating and maintaining the irrigation system, or in expanding it or making it more efficient, can be different from that of full-time farmers. Competing demands for labor time may interfere with their participating in certain management activities. This affects possibilities for widespread farmer participation.

(b) Where irrigation water has non-agricultural uses, significant conflicts of interest can arise -- for example, between farmers and fishermen (Abeyratne and Perera, 1984), or with operators of water-driven mills or turbines. To the extent that allocation decisions and investments in maintenance of supply must be interdependent, the "participation" of non-agriculturalists can crucially affect system performance.

(5) **Other Characteristics:** Other differentials like age or ethnic background may in some cases also be important to consider.

In most systems, there is scope for more productive user participation in the different activities of irrigation management, to arrive at decisions based on information from and agreement among those affected, to mobilize resources, improve coordination, and reduce conflict, thereby utilizing water supplies more advantageously. But how much more participation by whom will be beneficial, in what specific tasks and at what levels, will depend on the situation and problems at hand. Because there is some skepticism among various professionals about the feasibility of more participatory approaches to irrigation management, and because there is no basis for prescribing amounts or limits of farmer participation based on deductive rules, we present experience from many irrigation systems with respect to user involvement. This should provide a basis for enlarging water user roles in management where capabilities and incentives are appropriate.

Specialization in Participation

In irrigation systems developed and managed by users, one finds a wide range of situations, from minimal specialization -- where farmers participate in virtually all aspects of water management and any user may fill any roles that exist -- to a high degree of specialization -- where certain persons are assigned to very specific tasks. Consideration of how users have organized themselves illuminates some of the possibilities and problems for increased farmer participation in management.

Irrigators can evolve specialized roles for several different reasons: in the name of efficiency, according to the logic that division of labor leads to more output from given inputs of resources; in the name of expertise, as certain tasks may be done better by those with special skills and experience; or sometimes for the sake of status, as better-off farmers may either preempt or avoid certain responsibilities.[3]

There are a number of ways in which specialization can vary. First, there can be high or low degrees of specialization. In the Iranian qanat system described by Spooner (1974), cultivators did not have the technical skills needed to build or maintain the underground channels that brought their water. Craftsmen were hired to work on the qanats when in need of repair. At the other extreme, one finds the Marakwet furrow systems in Kenya, where the only specialists are the "blowers" who sound an alarm when they find a channel needing emergency repair. Everyone there works on the construction and repair of furrows as well as on operation and maintenance.

Specialist roles can be long-standing ones, appearing to be "traditional," or they can be recently introduced, more "modern" roles. The elders who oversee the Sonjo irrigation systems in Tanzania represent one extreme, while the officers in farmer organizations serving the Muda scheme in Malaysia can be taken to represent the other. The first type may be less amenable to deliberate revision of responsibilities, though Coward (1976) has documented the adaptation of a traditional water headman role (nai nam) to assume new functions of water management in a "modern" irrigation project in Laos.

A third way that specialization can vary is whether the roles are devoted entirely to irrigation management or handle other tasks as well. The responsibilities of members of the Juntas de Vigilancia de Riego in San Pedro de Atacama, Chile, are not so heavily specialized that only a few persons can discharge them, but one can say the roles are "specialized in water" (Lynch, 1978). In contrast, the shaykhs who handle irrigation tasks in the Daghara irrigation system in Iraq are tribal authorities with many other duties to discharge as well. Hunt and Hunt (1976) refer to this as "embeddedness" of irrigation roles.

While the overall level of specialization should not make much difference to government or donor agencies seeking to support farmer participation, it is probably true that gaining cooperation will be easier to the extent that existing specialist roles are more contemporary in origin and more focused on water. These can provide more accessible rationale and incentives for introducing changes that further specific irrigation objectives.

An exception would be where the "modern" roles are occupied by persons possibly from the community who are selected externally, without regard for competence or local support. This occurs where roles are filled on the basis of political patronage. It was seen, for instance, in Sri Lanka after 1978 when a new government appointed Cultivation Officers in place of elected Cultivation Committees to manage irrigation and other agricultural problems (Moore, 1979). For any irrigation role to be effective, it must have legitimacy in the eyes of those with whom it works. This could be seen in San Marcos, Peru, where an irrigation agency hired a vigilante to supervise water distribution for the La Huaylla main canal. The appointee was the son of a man with political connections in Lima. As he had no standing in the eyes of the irrigators, he and the users ignored one another and he became irrelevant (Barbara Lynch, personal communication).

Where "traditional" irrigation roles exist they should be examined first, to see what capacity for management is operative within the user community, and second, to know what activities need to be discharged on a regular or periodic basis for the system to function properly. The richness of special roles for irrigation management developed around the world is remarkable. Describing these roles, their responsibilities, their relationships with irrigators and officials, the arrangements for payment, etc. could easily fill a book, and the detail would be overwhelming.

To give readers an idea of various irrigation management roles that operate reasonably effectively in existing systems, we offer capsule descriptions from half the cases in Annex 3 (pages 168-174). Any reading of the documentation from which these profiles came should satisfy skeptics that water users have considerable capability to handle most irrigation tasks up to a fairly high level. Whether they will have the motivation to invest talent and energy is another matter which needs to be considered separately.

Incentives for Participation

The objectives analyzed in Chapter 2 encompass the major incentives that farmers would have to participate in various management functions. To the extent that opportunities like increased production, improved water distribution, or reductions in conflict are valued by users and can be promoted through their taking certain responsibilities, the prospects for user participation are increased. We consider here some of the main factors affecting such incentives.

Water Supply. The adequacy and reliability of water supply is often the main factor which influences user decisions to participate. Where water is abundant, there is little need for users to undertake tasks apart from maintenance, and in some cases, even maintenance work may not be necessary (Valera, 1985). At the other extreme, if water supply is too scarce or unreliable, and collective action cannot lead to any improvement in supply, participation offers little payoff.

This suggests a curvilinear relationship between water supply and incentives for participation, with negligible incentive at either extreme of scarcity or excess. Few empirical studies have tried to describe the "curve" this implies. Uphoff et al. (1981) suggest that the relationship is shaped like an inverted U, with user's willingness to invest in participation being low at either extreme of water abundance or scarcity. Farmer opinion data from the Gal Oya scheme in Sri Lanka support this hypothesis statistically. Wade (1984), drawing on his study of 31 villages in the Indian state of Andhra Pradesh, proposes a slightly different shape, an inverted J. More systematic research would be required to establish such a relationship more precisely, but some empirical work supports the proposition that "participation" would be most active and extensive in some middle range of water availability.

The exceptions would be where collective action could increase supply -- by constructing and managing facilities for acquisition or by increasing water allocation and distribution from higher levels within large systems -- or could curb damage caused by excess water. As suggested in Chapter 2, since participation entails invariably some costs for water users, there is no basis for trying to maximize farmer involvement in irrigation management activities. Rather some "optimum" should be sought. The "middle range" within which participation provides net benefits to water users may be quite broad and may be hard to delimit quantitatively, however.

Cost or Difficulty of Water Acquisition. Factors besides gross supply can enter into the calculus for user participation. Studies of hill irrigation systems in Nepal found that roles and rules were more formal and specific where greater (but not inordinate) effort and expense were required to assure water supply for the commanded area. Organizations tended to be more elaborate in terms of recording minutes of meetings, keeping track of work contributions, etc. where the terrain was unstable or presented great impediments, where diversion structures to capture water required much labor to rebuild or repair, or where tunnels had to be built to convey water. Farmers' investment of effort in organizational activities was correlated with their need to invest in activities concerning water and the associated structures, physical and jurisdictional, necessary for obtaining and conveying it (Martin, 1986: Ch. 5). This suggests that difficulty in acquiring water, more than its relative scarcity, will determine how extensive is users' organization for irrigation management.

This proposition is consistent with an analysis of farmer participation "above the outlet" (Chambers, 1984) where the main impetus for such activity seems to be the need to obtain and assure adequate water. As noted previously, acquisition may be accomplished by group labor to capture water and construct channels, or it may be a matter of lobbying to get favorable allocation decisions made at higher levels. Sometimes collective action may be needed to ensure -- by patrols, by bribes, or even by threats or use of force -- that allocated water actually gets delivered, as seen from the Andhra Pradesh and Tamil Nadu profiles in Annex 3.

Location. Spatial factors often affect who participates in irrigation management and who occupies positions of leadership. The traditional leaders who direct the ahar-pyne systems in Bihar state of India on a very informal basis have an incentive to get the water distributed broadly and fairly because they, like everyone else, have landholdings in all parts of the command area, due to high fragmentation of holdings. The same phenomenon is reported in the Mexican systems Downing (1974) studied. In Andhra Pradesh state, Wade (1979) found that community organization for irrigation did not arise or function well in villages where the largest and most influential landowners had their holdings in head areas (near the canal) and thus had no incentive to support farmer organization. If such persons had their land at the tail, or if holdings were more equally distributed, community organization for irrigation was more likely to emerge and be effective. We have noted the general lack of such organization in the Punjab state of Pakistan, but Merrey (1983) found that where large landowners had problems with water supply, they were able to mobilize work groups for necessary maintenance activities.

It is important to consider what other incentives farmers may have to participate actively and equitably in water management efforts, collectively and individually, either as leaders or as members of organizations. Considerations like the scarcity or difficulty of securing water supply can be significant motivating factors. They are not the only ones, however. While the self-interested concern of getting water for one's own field is a real and urgent influence on behavior, the extent of collective action is seldom fully explained by purely material considerations.[4]

One commonly finds farmers placing some value on maintaining harmonious relations within the community of water users and on ensuring users at least some minimum of subsistence. To the extent that access to water dramatically affects people's life chances and even survival in some circumstances, there are ethical elements attached to sharing water which can mobilize cooperative behavior among irrigators more readily than among rainfed agriculturalists, who are less interdependent in their mode of production (Uphoff and Van Dusen, 1984:41-46). In assessing possibilities and plans for farmer participation, the existence and balance of material inducements for entering into water management activities need to be considered. But narrow "materialist" or "individualist" views should not be allowed to obscure the possible incentive effect of non-material or shared group interests.

Mobilizing Leadership for Participation

Participation of any kind stems from people's decisions to devote a portion of their time, thought and energy to dealing with problems through some form of collective action. Organization makes participation patterned and predictable enough to acquire some recognizable and productive structure. Incentives, discussed above, give people motivation and make participation more sustainable. A third influence is leadership, which makes participation more coordinated and effective by providing direction, encouragement and discipline. This very complicated subject is not treated adequately in social science theory, but some discussion is in order here because of its significance for farmer organization and participation. The effectiveness of organization and the sustainability of participation depend crucially on the quality of leadership attracted from among water users.

The function of leadership is to plan and carry out decision-making, resource mobilization and management, communication, and conflict management, though not necessarily to implement them all personally. By taking initiative and responsibility, by coming up with strategies and concrete proposals, by talking with, persuading and possibly disciplining others, persons in leadership roles can energize groups to achieve at least part of the potential that rests in their aggregated capabilities, interests and ideals.

Roles of responsibility for irrigation management are filled in various ways. In the traditional irrigation cases from Tanzania and Oman, leaders moved into their positions by inheritance, whereas at the other extreme, in the irrigation groups in Zapotec, Mexico, leadership roles are filled by rotation.[5] More commonly, irrigation leaders are chosen through consensus or election from the community of water users, or through selection by higher authorities.

In systems that are largely agency-managed, or where there is a large agency role, persons in management roles at lower levels may be chosen or appointed by officials even if the persons selected are nominally responsible to users. In some of the Philippine systems under the National Irrigation Administration where farmer organizations are being introduced, farmer leaders are sometimes selected by NIA's group organizers, while others are directly elected by farmers; still others are chosen by a process in which the organizers participate (Illo and Chiong-Javier, 1983). In the Pochampad scheme in India, the representatives of farmers who constituted Pipe Committees were at the outset chosen by agency staff, but increasingly these choices were left to farmers. The consequence was that "real" leaders were more often brought into irrigation management, with better results according to Singh (1982).

Many cases suggest that mobilizing new leadership and talent into positions of responsibility is crucial for improving water management. There is much to be said in favor of working with and through "traditional" leaders, as done for example in the Laos and Iraq cases. But water management requires both special skills and strong commitment for high performance. If existing local leadership is satisfied with the status quo or tied into political networks outside the community, there is reason to look for new persons who have the aptitude and interest for improving irrigation.

We saw in the preceding section how "traditional" leaders in India and Mexico worked to ensure reliable and equitable water availability, at least partly because of incentives created by patterns of land fragmentation or allocation. On the other hand, the mobilization of new leadership through deliberate government initiatives has energized water management in the Pochampad scheme in India, in the Gal Oya and Minipe cases in Sri Lanka, and in many systems "organized" by the National Irrigation Administration in the Philippines. There is no evident basis for always preferring to work with either "traditional" or "modern" leaders in promoting more participatory water management. One finds in the literature both positive and negative experience with both kinds of leadership. The key factor seems to be ensuring "accountability" as discussed by Coward (1976 and 1977) and as taken up in Chapter 8.

User Groups vs. Local Government

The most common institutions for handling water management responsibilities at the local level are user groups, often referred to generically as "water user associations" (Cernea, 1984). On the other hand, one finds some instances where decisions about irrigation, resource mobilization and management, communication and conflict resolution come under the authority of local governments. These can range in their origin and style from a council of hereditary elders, as with the Sonjo in Tanzania, to the "modern" Village Council arrangement established in the Rahad scheme in Sudan.[6]

The traditional system of water management in Daudzai, Pakistan appears to function quite well through a village council of elders (mashers). Yet the notably effective traditional subak systems in Bali, Indonesia operate separately from the village (banjar) authorities. When the Indonesian government introduced new models of irrigation organization in Java, it kept the traditional water headman role (ulu-ulu) but did not tie it into the village authority structure as had been the case traditionally. The village headman (lurah) is now sometimes a sponsor of the dharma tirta irrigation organization at the village level but in some other locations he has no role (Duewel, 1984).

In Oaxaca, Mexico, it was previously the case that irrigation groups were linked to the community government, and persons became eligible for leadership roles in the tramos by having taken on community responsibilities such as sponsoring fiestas. However, as the status and authority of these traditional civil-religious roles diminished, this link has become attenuated, as has community control over irrigation. The expansion of the state apparatus is undercutting both traditional local government and community resource management. In Peru, the abolition of traditional civil-religious roles greatly detracted from local capacity for irrigation management (Mitchell, 1976).[7]

That traditional village authorities in Andhra Pradesh, India remain effective in resource management tasks is encouraging, the more so because this local capacity seems to be demonstrated with most vitality in communities where it is most needed (Wade, 1984a).[8] It might have been expected that the more "modern" local government structure of the panchayats would have displaced the indigenous system of elders in handling organizational activities for irrigation. While Wade did not find all communities discharging such functions effectively, there was evidence of reasonably good performance in raising funds, hiring specialized staff to guard the water and fields, and upgrading facilities. Performance could have been still better if the agency managing the main system would have cooperated with the communities more fully and fairly.[9]

One can make arguments for channeling farmer participation in irrigation management either through user groups or through local government bodies. Both kinds of institutions have registered some impressive successes, and both have given their share of disappointments. In our comparative analysis of local institutional development, we noted that local governments throughout the

Third World have generally shied away from responsibilities for agriculture (Uphoff, 1986a: Chapter 5).

User groups appear to be generally a more promising avenue for irrigation management than are local governments. Because both the set of resource users and the water resource itself are relatively definite and delimitable, there is basis for gaining users' cooperation and compliance through processes of discussion and consensus, rather than needing to invoke "authority" as local governments are able and even prone to do (Uphoff, 1986a: Chapter 2). Even if user groups play the most active role in water management, they are likely to need links with local governments which usually will have some role in dealing with land matters.[10] Often some combination of membership and local government institutional channels will be most favorable, combining elements of voluntarism and authoritative action to derive common benefits.

Participation of Women

Our review of the literature revealed very little documentation on participation by women in irrigation management. In San Pedro de Atacama, Chile, women constitute about one-third of the landowners, given high male out-migration, yet their role in the irrigation organization is minor. Women can be ditchtenders and sometimes members of the Juntas de Vigilancia de Riego, but in general their role is restricted to participation at meetings. In the Mwea scheme in Kenya, there are informal women's work groups active in irrigated cultivation, but women are excluded from public meetings (barazas) with officials. The Rahad project in Sudan has taken the unusual step of reserving six seats on each Village Council for women, but these councils are relatively unempowered. While one might expect women's roles to be negligible in an Islamic country, one of the few documented activities for women in irrigation is in the Nayband irrigation system in Iran, where women time the rotation of water deliveries.

These are the most specific references to women's participation that we found in our analysis of the 50 case studies. We know that women are almost always heavily involved in the agricultural activities associated with irrigation. Yet we do not see much direct women's participation in managing the water, structures and processes whereby agriculture is made more productive through irrigation.[11] Scholars and practitioners need to be more attentive to the actual and potential contribution of women's participation in irrigation management.

NOTES

[1]In certain schemes, tail-enders, or at least middle-reach cultivators, may be better off than head-enders due to soil, topographic water distribution or other factors. So the general rule should not be overstated or stereotyped, even if it is a rule (Wade, 1980; Moore, 1983).

[2]The Sonjo in Tanzania have four categories of households, some with full water rights which are inherited within the family, some with limited water rights, some which must pay for monthly water allocations if they are to receive any water, and the remaining households which may be "clients" of other households and get any surplus water not used by their patrons. Heads of households in the first category make up the council which makes all decision governing the irrigation system as well as other community affairs. The second have membership in the council but no voting rights. The other categories must participate in maintenance work but have no voice in the running of the system (Gray, 1963).

[3]This latter explanation seems to be more important than the other two, for example, in the Cumbum valley, Tamil Nadu state in India where wealthier farmers prefer to concentrate on agriculture or other activities, assigning irrigation tasks to less well-off persons. (Ramachandran, 1984). Irrigation specialists such as ditchtenders, often low-caste, appear to be more common in communities that are more highly stratified socio-economically. Such arrangements are reported in India (Wade, 1979; Meinzen-Dick, 1984) and also Indonesia (Geertz, 1967), where specialist groups of poorer farmers are paid to handle distribution and maintenance (see footnote 10 in Chapter 4).

[4]An analysis of the progress in introducing participatory water management organizations into the Gal Oya scheme in Sri Lanka concludes that farmer motivation should not be seen as limited only to material rewards (minus material costs) or only to individual net gains (Uphoff, 1985a). Some value is often placed on the well-being of others, whose problems and benefits become more salient once organization is created, though probably not to the extent that others' gains come at one's own expense in zero-sum terms. Concern for others' benefit is generally manifested within the parameters of what economists refer to as "Pareto optimality."

[5]The wakils (agents) who supervise the Izki irrigation system in Oman are chosen from certain families that are thought to be

"above" the tribal passions and conflicts that otherwise divide the community (Wilkinson, 1977). The practice of assigning leadership responsibilities by strict rotation is reported also for traditional Japanese communities at the small group (kumi) level, where responsibilities often included irrigation (Sasaki, 1985).

[6]This latter system is so heavily permeated by agency control that is is not performing irrigation management tasks very satisfactorily. The USAID evaluation reports that the government seems fairly serious now about permitting more local responsibility, so this might become a more effective system in the future (Benedict et al., 1981). It was modelled after the Gezira and Mwea schemes.

[7]This is described briefly in Annex 3. A different kind of decline in local capacity for irrigation management in Peru is reported by Barbara Lynch (personal communication). On the La Huaylla main canal in San Marcos until about 1970, the main water management role was the water judge. These judges were usually millers, whose legitimacy and authority were based on (1) their strong interest in maintaining good water flow to their mills, (2) their disinterest in the ultimate distribution of the water among farmers, and (3) the fact that they were water users but not water consumers. However in recent years the status of the millers has declined, the demands for water have increased dramatically, and a major flood destroyed much of the old physical infrastructure. The declining authority of the old water judges (millers) was exposed by the changing conditions and was found to be insufficient to survive.

[8]Recall the reference on page 84 above to a "J-shaped" curve proposed by Wade relating water availability (or feasibility in procurement) to participation levels. This suggests that below some level of availability (or above some level of difficulty) participation drops quickly to zero. This may be because some amount of participation -- some threshold -- needs to be surpassed before participation is effective and can be self-sustaining.

[9]An impediment to community management is the institutionalized system of corruption governing water allocation and distribution, documented by Wade (1982a and 1984). One of the intriguing "resource mobilization" mechanisms controlled by councils of elders is to re-auction the license for selling liquor in the village, reported in the Indian states of Andhra Pradesh (Wade, 1979) and Tamil Nadu (Palanisami and Easter, 1983). The license is officially auctioned off by the government to raise revenue for itself, but the villages regard licensing as their prerogative and as a legitimate source of income for local purposes. They contrive to

rig the bidding by boycotting or otherwise frightening off any outside bidders. Someone from the council buys the license for a low price at the government's auction and it is then re-sold in a competitive village auction that raises money for common activities such as paying water guards.

[10]In the Andhra Pradesh systems Wade studied, where traditional local governments oversee water management, the non-irrigation functions performed by these village organizations are probably important for reinforcing the authority they exercise over water (personal communication).

[11]In her report on observations of women in Asian irrigation, Cloud (1982:2-3) identified a fairly common role for women as informal negotiators in conflict resolution, where they could and did reconcile disputes between male irrigators.

6
The Context of Participation:
Factors in the Environment

The possibilities and productivity of farmer participation will be conditioned by the situation in which water users find themselves. The incentives and constraints they face derive from many sources, but three broad categories of factors appear most important:

(a) historical factors that affect farmers' willingness and ability to assume responsibility for various irrigation tasks,

(b) physical and economic factors that shape the supply of and demand for water, thereby affecting water users' orientation toward collective action, and

(c) socio-cultural and political factors that influence the way water users relate to one another and to the government.

Each of these areas could be analyzed in a chapter by itself, but it will suffice to review some of the effects which context can have on the nature and extent of farmer participation for irrigation management. This should help to guard against sweeping generalizations or uniform policy pronouncements that ignore contextual differences.

Historical Factors

The knowledge and skills which water users can bring to the tasks of management, as well as their disposition to accept responsibility, will depend a great deal on the origins of each irrigation system. The extent of farmers' knowledge of irrigation and their knowledge of the area can be summarized in four alternative situations, outlined in the matrix below.

Farmers will generally be able to make the greatest contribution where they have experience both with irrigation and with the area (I). This includes not only contributions to operation and maintenance, but also to design and construction where a system is being rehabilitated or expanded.

Where irrigation is being brought to a rainfed cultivation area, some technical assistance or training in water management principles and techniques may initially be necessary (II). This was seen in the irrigated perimeters along the Senegal River where such efforts were made by the administrative agency. The same point can be made through a negative example in the case of the Pultan Para pump irrigation system in Bangladesh where a lack of technical knowledge by users gave rise to inefficiencies of operation and to ensuing conflict.

Figure 3: WATER USER KNOWLEDGE BASE FOR IRRIGATION MANAGEMENT

FARMERS' KNOWLEDGE OF THE AREA	FARMERS' KNOWLEDGE OF IRRIGATION: High	FARMERS' KNOWLEDGE OF IRRIGATION: Low
High	I: Farmers in traditional irrigation system (e.g. Chhatis Mauja, Subaks) or in long-established irrigation system (e.g. Abu Raya)	II: Irrigation being introduced into settled farming area (e.g. Matam, Bakel, Pultan Para)
Low	III: Resettlement of farmers familiar with irrigation into new irrigation system (e.g. Muda, Gal Oya)	IV: Resettlement of rainfed farmers in irrigation system (e.g. Mwea)

To be sure, rainfed farmers in an area may have some important information such as that pertaining to local topography or soils, and traditional forms of social organization may be adaptable for irrigation management. Formerly rainfed farmers or pastoralists may require some training about land leveling or irrigation practices in order to make the most useful contributions from their existing knowledge. An intermediate case between I and II would be where farmers have been irrigating but a new technology such as mechanized pumps is introduced. In such instances, training is very important to give command over the new technology, but it would not be a matter of versing them in basic irrigation skills.

Farmers in resettlement projects cannot be expected to have detailed local knowledge, at least at the outset. If they come from irrigated areas (III), they may have enough skill in water management to be able to acquire quickly and to utilize information about the new system and its environs. Resettled farmers who lack experience with irrigation require the most training and assistance before they can engage in management activities (IV). Moris and Thom point out that in large African irrigation schemes:

> Where initial extension and training has been weak, farmers become entirely dependent on scheme management for advice, and they are likely to perform certain key operations (like field leveling) so poorly that yields are greatly depressed. This situation in turn reinforces stereotypes held by managers and staff about farmers' low motivation and interest... (1985:79)

Large irrigation and resettlement projects often draw farmers with varying degrees of experience in irrigation, and one of the difficulties in getting them to assume responsibility effectively can be the heterogeneity of skill and will among water users for participating in management of the system. An agency working in an irrigated settlement area should involve farmers in planning its training and technical assistance so that appropriately differentiated kinds and amounts can be provided.

These considerations speak mostly to farmers' ability to handle water tasks proficiently. There are also factors affecting their willingness. The pattern of prior investment, by creating concepts and claims of "property," will have an effect on water rights and responsibilities for system performance (Coward, 1983). To know who made the initial decisions and investments to create the irrigation system indicates in whose interest the system was conceived, located, designed and built, whether more according to farmer or to government needs. To observe this is not to suggest that governments do not wish to look out for farmers' interests. But the way it assesses them is often different from the way farmers perceive their own interests. Governments are likely to think in terms of raising national production, whereas farmers are more concerned with the returns to their land or labor. Governments may be looking for projects that can win donor funding while farmers are seeking to meet family income needs.

When users make the original decision and then construct an irrigation system, they have more identification with it and are more inclined to take an interest in operation and maintenance. In community systems in Nepal, membership and rights to water are based on investment either in the initial construction or later extension of the physical facilities. For instance, the history of previous community investment in the Tallo Kulo scheme presented certain constraints when negotiations were undertaken with the government for expansion of the system. Similarly in some of the zanjeras in the Philippines, the pattern of prior resource contributions created two categories of membership with different responsibilities. Those who had provided the land that became irrigated did not have to provide their labor for subsequent maintenance work, as did other members who built the system with their labor in order to secure access to irrigated land.[1]

In some systems such as San Pedro de Atacama in Chile, it may be necessary to take into account any existing claims of non-agricultural users of water traditionally used for irrigation -- e.g., for copper mining or urban domestic use. The consequences of having competing non-agricultural uses are difficult to predict for water user associations. Conflict with other users, especially if they are politically powerful, may make cultivators' control of water so unreliable or insecure that their disposition to sustain collective action is undermined. On the other hand, an outside challenge to their subsistence can contribute to group solidarity. One can only note that precedents of non-agricultural use of irrigation water need to be considered. Some understanding of what has gone before is important for appreciating who will be willing and able to participate in irrigation management and in what manner.

It is also important to apply an historical perspective to the irrigation system itself, in order to recognize that it is continually changing along various dimensions, such as water availability, intensity of use, population pressure, cropping patterns, market conditions affecting profitability of production, etc. The Greek philosopher Heraclites, arguing an epistemological point almost 2500 years ago, achieved immortality with the observation that one can never step in the same river twice -- it is always changing, if only because one has stepped in it. The same can be said of an irrigation system. "History" encompasses not only an awareness of antecedents but also of continuous change. The continually changing factors noted here need to be considered in terms of their implications for farmer participation, particularly contextual factors which affect the supply and demand for irrigation water, examined next.

Physical and Economic Factors

These might appear to correspond respectively to the factors which affect supply of and demand for irrigation water. But in practice no definite dichotomy can be found since both factors are interactive. Moreover, both are affected by socio-cultural factors, discussed in the following section, so a distinction between physical and economic factors is expositional more than explanatory. In the real world, farmers' dependence on irrigation, which gives them a stake in managing it effectively, is established in part by the natural environment (climate, topography, etc.) but also by the absence of alternative economic opportunities. Cropping patterns and agronomic practices can be regarded as either physical or economic. There is a continuum of contextual factors from clearly physical (like soil and topography) to essentially economic (e.g. availability of labor). The whole set of factors bear on relative water scarcity, a concept which reflects the balance of supply and demand.

These factors are important because of the observation, offered in Chapter 5, that farmers' investment of time and effort in irrigation management activities reflects their need for adequate and reliable water. Where water is abundant, there may be little or no need for collective effort to acquire, allocate and distribute it (though there may be need for drainage activities under certain natural conditions of soil and topography). Water scarcity gives impetus to farmer organization and participation, but we also observe that where the supply is quite limited or unpredictable, the return from farmers' investment of resources in irrigation facilities and activities can be too little or too risky. Both supply and demand are relative. A very small supply might be adequate for a few farmers if it could be obtained at an acceptable cost. Physical considerations verge invariably on economic and even social and ethical concerns. The value attached to even meager or unreliable supplies of water can be very great.

Factors Relating to Water Supply. One can consider supply of water for irrigation in a gross physical sense. Is the water that flows in a river, that can be captured in a reservoir from rainfall run-off, or that is available in underground aquifers for pumping (a) sufficient in amount and regularity, and (b) sufficiently accessible for exploitation in irrigation? The first consideration of sufficiency is affected by patterns of rainfall, topography and soils, and are relatively fixed. But the second is often quite variable, leading into questions of alternative technologies and into the availability and cost of factors of production (capital and labor) which can be invested to capture and convey the water. Technology itself is not just physical, as the feasibility and use of any particular technique for acquiring water depends on skills and organization.

The source of irrigation water affects what technology is appropriate and what kinds of skills and organization are needed. Sources are usually classified as (a) surface, or (b) groundwater, with the first relying mostly on gravity-flow techniques and the latter requiring some kind of lifting arrangement. In fact, the first may be augmented by lifting, and the latter almost always uses gravity techniques for distribution.[2] The nature of the technology used certainly affects what specific tasks must be undertaken to manage the physical system. It also affects the degree of specialization of roles in carrying out these tasks. For example, pump schemes and the qanats discussed previously require considerable expertise to construct and repair, so these tasks are generally done by specialists, often from outside the local community (Spooner, 1974; Wilkinson, 1977; Sutton, 1984). Users need to be organized only to the extent that they can mobilize the financial resources to acquire such services. In contrast, many surface irrigation systems use large amounts of relatively unspecialized labor for construction and maintenance.[3]

In an irrigation system, topography and soils affect what tasks must be accomplished and the amount of work involved. Surface gravity systems in flat, heavy soils will require more attention to drainage structures and activities than in steep hilly areas with light soils. In the latter situation, however, frequent landslides are likely to direct farmer efforts toward patrolling and maintaining channels during the rainy season. This may require fairly elaborate organizational measures, as found in the hill irrigation systems in Nepal.

The scale of the irrigation system in terms of command area or the number of farmers involved affects the way in which management tasks are performed. Organizational tasks, in particular communication, become more difficult as the distances involved and number of people increase, and more formal mechanisms are generally required in larger systems. As shown in Chapter 4, however, even large systems disaggregate into smaller units of operation and of organization. The context of irrigation can be altered by organizational measures that mitigate the negative features of large scale by establishing multiple capacities for management within the system, making it a composite of many smaller systems.

Where there are variations in the supply of water, between "wet" and "dry" seasons or year to year, some combination of technological and organizational means needs to be found that minimize or at least compensate for fluctuations. Farmers' organizations often change their activities and mode of operating between seasons (or even within seasons) in response to variation in water supply (e.g., Downing, 1974). Government agencies, which tend to operate according to the calendar or fiscal year rather than the climatic year, need to make allowances for the different rhythms and patterns of seasonal variation in irrigation and not have the same expectations of water user associations at all times, such as holding meetings on the same day of every month.[4]

Factors Relating to Water Demand. Although fluctuations in water availability are evidently physical, they can be influenced by social activity through technological means. "Scarcity" results from the interaction of supply and demand. How much water is adequate, and how much effort at control is needed will depend on the demand for irrigation water, the uses to which it may be put and the benefits to be derived therefrom. Some factors influencing this are discussed below.

Cropping pattern. The crops being grown, their water requirements and the timing of application all affect water management needs.[5] Diversified cropping patterns and staggered planting dates in a command area may reduce peaks of demand, making "supply" more satisfactory, but this generally requires more careful management of water. Many coordinated deliveries are needed instead of fewer, larger amounts. Detailed management of demand through cropping patterns and through selection and timing of crops in a command area is an effective complement to water management strategies of farmer organizations in the case studies from Bali and Java in Indonesia. Similarly, many large-scale irrigation systems in India, and Pakistan, in Africa (e.g. Tono, Sabi, Mwea, Gezira) and in Brazil have sought to impose certain cropping patterns in the command area, with varying degrees of success. The objective of such regulation is generally to achieve management patterns that are simpler or more "efficient" from the agency's point of view. In some Indian schemes, zoning of crops with low water requirements has been a means of extending irrigation (and drought protection) over as large an area as possible. Cropping pattern, whether agency-regulated or farmer-preferred, will definitely affect user incentives and practices for irrigation system management.

Agronomic practices. The willingness of farmers to adopt water-saving techniques should not be assumed, particularly when these techniques require extra labor or financial cost. Farmers may well prefer to have continuous standing water to control weeds, for example, instead of using herbicides or doing more weeding by hand (Svendsen, 1983). Willingness to undertake such costs is affected in part by social or cultural influences which encourage certain "standard" practices, but particularly by the profitability of production, which needs to be viewed as part of the "context" of irrigation management.

Profitability of irrigated farming. This depends on the prevailing prices of crops and inputs as much as on physical levels of production. While many factors affect demand for irrigation, the cost of water itself, especially if priced volumetrically or by timed deliveries, may have the most direct impact on demand. In practice, it is rare to find strict payment for measured volumes, even though this is often advocated by economists. Farmers' "cost" of water most often is reckoned in terms of the time, effort and money users need to expend to ensure their supply.

The value to farmers of engaging in more detailed irrigation management is affected both by economic and agronomic factors and by social and cultural considerations. The profitability of production (or the need and desirability of subsistence crops) will determine farmers' return from alternative cropping patterns, which in turn will influence demand for water and affect its relative scarcity.

It should not be surprising, for instance, that there is little constructive farmer participation in irrigation management in the Chaj Doab village studied by Merrey (1983 and 1984). He calculates that the 'fund of rent' that has been extracted through land taxes, irrigation fees, rent to landlords, unofficial payments (to officials), and unfavorable terms of trade has made villagers worse off than before the introduction of irrigation. In such conditions, there is no incentive for "optimal" investment of farmer resources in irrigation management to push yields to their highest efficient level, all input costs considered. Rather, water users make the minimum investment required to obtain subsistence production levels. More effort would only enrich others, not the farmers and their families.

Dependence on irrigation. Where large portions of a community rely on irrigated subsistence or commercial production for their livelihoods, active responses to shortfalls in supply are to be expected. Dependence on irrigation, and hence the incentive for participation in water management is lower where irrigation is only supplemental to rainfall, where rainfed cultivation is an important source of food or income, or where farm families have other major sources of income.

Most farmers in the dry zone of Sri Lanka depend heavily on the output of their rainfed shifting cultivation (<u>chena</u>) since with small irrigated holdings, their production of rice is not sufficient for family needs (Leach, 1961). This continually frustrates irrigation and agricultural officials who expect farmers to give priority to their irrigated rice. Farmers often are away in upland areas planting their chena crop right after the first rains when officials want them to be preparing their irrigated land in unison to make simultaneous water issues possible; or farmers refuse to do maintenance at certain times when they need to be working on their chena because they expect more income from it than from a somewhat improved rice crop.

In the Mwea case in Kenya, the agency sought to restrict the size of unirrigated gardens and supplemental businesses in order to get tenants to work harder on their irrigated plots. However, rice was primarily a cash crop, while traditional unirrigated crops were grown for family consumption. Moreover, the price the government paid farmers for their rice was below the market price, so irrigated farming appeared unprofitable.

Availability of labor. This is closely related to dependence on irrigation and to the profitability of production, but it should be considered in its own right, since it is a complex variable. One can say in general that in Asian communities where labor is abundant relative to land and other resources, there are lower opportunity costs for labor than in an African context where it is relatively scarce. Households in labor-scarce areas may be better rewarded by an extensive rather than an intensive production strategy.

Where other economic activities are important, labor shortages for irrigation-related activities are more common. Yet, even where irrigated farming is a primary source of food or income, there can be shortages of labor for some tasks. Migration may have drawn away part of the labor force, as in San Pedro de Atacama, Chile, or because households allocate part of their family labor to upland or garden crops, as among the Sonjo in Tanzania. Labor shortages are especially noted in areas of Taiwan and South Korea, where there are large numbers of part-time farmers.[6]

Irrigation's demand for labor is subject to considerable variation within and between seasons, so labor constraints on irrigation management activities can occur at certain peak times even if the overall supply of labor appears sufficient. The timing of irrigation-related activities and the expected level of farmers' labor contributions must take seasonal and average labor shortages into account.

Land tenure. Land arrangements can affect both the willingness and ability of water users to participate in irrigation system management, as noted in Chapter 5. For example, tenants with an insecure claim to the land they cultivate are usually less willing than landowners to contribute to permanent capital improvements in the irrigation system. They may also be less able to make investments because of a lack of credit available without land to pledge as security.

Where landholdings are quite unequally distributed, the manner in which farmers participate in irrigation management will be influenced by the resulting power differentials, as in parts of Pakistan (Merrey, 1983). There can be tenure-based differences even in ostensibly more egalitarian organizations like the subaks in Indonesia.[7] In several Nepal cases, it was reported that farmers' investments in upgrading the irrigation system and their management activities increased after land reform was introduced in that country.

Degree of commercialization. Although many of the economic factors mentioned here apply to both subsistence and cash-cropping areas, this variable can have an important bearing on resource mobilization for irrigated agriculture. Farmers engaged in subsistence production in a less-developed cash economy will have to rely primarily on labor and in-kind contributions for financing construction as well as O&M responsibilities. Cash is likely to replace material contributions in more commercialized areas, and may eventually replace labor contributions where there are wage labor opportunities, especially if there is a shortage of family labor, as has been occurring in Taiwan.

Economic explanations are not sufficient by themselves, however, to predict irrigation management behavior. One might expect farmers to "monetize" their contributions to irrigation management more where commercialization of agriculture is greater. There is little apparent difference between the South Indian state of Tamil Nadu and Sri Lanka in terms of the level of commercialization. Yet case studies indicate a much greater disposition of cultivators to hire laborers to do O&M work on their behalf in Tamil Nadu (Meinzen-Dick, 1984; Ramachandran, 1984) than in Sri Lanka (Leach, 1961; Widanapathirana, 1984). In part this could be due to a greater surplus of labor in Tamil Nadu and a lower real cost of employing workers there. But practices appear to be influenced also by differences in caste structure and community norms. This calls our attention to social, cultural and other variables in the context of irrigation management.

Socio-Cultural and Political Factors

The contextual factors reviewed so far are relatively observable and measurable. However, the influence of more abstract institutions, values and ideologies in shaping farmer orientation toward participation and organization should not be underestimated. We have already considered in the preceding chapter the significance of indigenous forms of leadership and social organization as providing a basis for cooperation among water users.

Ethnic and other social differences can form lines of cleavage and potential conflict among water users, especially if these correspond to differences in land tenure status or in access to water.[8] Fortunately, we find ethnic and social solidarity within groups of cultivators often contributing positively to their willingness and ability to work cooperatively, as seen with the zanjeras in the Philippines or the Marakwet case in Kenya.[9]

There may be culturally sanctioned forms of conflict management, such as those deriving from Islamic law, which help in the operation of water user groups (Fernea, 1970; Wilkinson, 1977). On the other hand, cultural values do not always favor cooperation. The concept of izzat (honor) is a source of considerable strife in the Pakistani community studied by Merrey (1982), and charges of witchcraft based on traditional beliefs kept the two irrigating wards of Wamira in Papua-New Guinea divided through much of their history (Kahn, 1983).

Just as cultural ideology affects the degree of cooperation among farmers, political ideology can have an equally important influence on cooperation between farmers and an irrigation agency. One of the most striking examples of this is seen in the Meichuan system in China, in which the agency technical staff worked alongside farmers during a crisis to gain their confidence, and adapted the farmers' system of small reservoirs and ponds to make an interconnected "melons on a vine" system with much greater capacity and flexibility for irrigation. Such strong populist sentiments are rarely found or put into practice in government agencies, particularly where the technical staff have considerably higher levels of education and social status than farmers. Still, a political ethos which is more favorable toward participatory methods and egalitarian outcomes will make it easier for engineers and administrators to work more cooperatively with water users.[10] These considerations of political support for farmer participation and improving the working relations between officials and water users are taken up in the next chapter and Chapter 9.

* * * * *

This consideration of the context of farmer organization and participation for improving irrigation management concludes Part II of our analysis, which lays out the main dimensions along which tasks, opportunities, responsibilities and conditions for organization and participation may vary. While irrigation is an immensely complex undertaking for both water users and system managers, the analysis presented here integrates and simplifies the enterprise conceptually. In the final three chapters we discuss the policy, design and administrative implications of a participatory approach to irrigation management that encompasses the interests and talents as well as the resources of water users.

NOTES

[1]This illustrates the value of distinctions between kinds of irrigation activities as introduced in Chapter 3, between those directed to the physical structures and those contributing to the organization. In the creation of the system (design and construction), persons who owned unirrigated but irrigable land pooled this resource with persons who had only their labor, who could build the facilities needed for irrigation. In this cooperative enterprise, the new asset created (irrigated land) was divided among organizational members, who had, however, different recurrent obligations in O&M.

[2]One possible but expensive technology is for water to be pumped directly to each farmer's fields, as done in the Battar irrigation project in Nepal as part of a World Bank integrated rural development project there. The project appraisal report said proudly of the designed system that "positive water control was achieved, and distribution losses effectively eliminated, by the installation of a piped field distribution system." Unfortunately, the system of piping water to the fields did not allow the cold water being pumped from the river to warm in the sun as happens in a gravity system when farmers bring water to their fields by surface channels. Farmers complained that their crops did not do as well because of the difference in water temperature. They suggested that had they been consulted on the design, they could have warned against this very costly and impractical technology. A further problem was that the government gave priority to urban centers in the allocation of electricity so farmers suffered from frequent interruptions in water delivery (Uphoff, 1985b:366).

A variant on this technology is the "demand scheduling" system being experimented with in the Mahaweli "H" system in Sri Lanka, where each field has a pipeline with a control valve so that farmers can get water any time they want it and in whatever amount they think they need. As in the Battar scheme, this is intended in part to obviate the need for farmer organization and participation. While this terminology does not present water temperature problems, there are other problems such as breakage and wasteage that will have to be eliminated to justify the cost, almost twice as much per acre as for conventional gravity distribution (Gunston, 1983).

[3]The extreme example would be the Marakwet furrow system in Kenya where everyone contributes to all activities, the only special roles being for "blowers". Gravity-flow systems constructed in hilly regions often require tunnelling to carry channels through as well as along mountainsides. Some amount of "expertise" for this work may be mobilized from outside the community, as in the building of the Chherlung system in Nepal (Martin and Yoder, 1983).

[4]For an agency, the ebbs and flows of activity associated with the fiscal year are similar to those that farmers have for a cropping year, with peaks and troughs of work arising as the scarcity or abundance of the critical resource (money/water) influences patterns of activity (even attitudes and tempers). We are not dealing with the context of agencies' performance or more would have to be said on this matter, which farmers are as likely to misunderstand or overlook as officials are likely to ignore seasonal variations.

[5]Some data on different crop water requirements are given in Seckler (1985). His argument for letting paddy-cropped irrigation systems operate with no agency involvement is controversial and not widely accepted, as seen in the responses to his paper in the ODI network paper.

[6]It should not be assumed that farmers with small holdings necessarily have extra labor. They may be more involved than other farmers in supplemental subsistence activities such as crafts and wage labor.

[7]A distinction is made in membership status and obligations according to land tenure status. Those who are landowners but not cultivators are subak members for purposes of decision-making, but do not participate in the teams (pekaseh subaks) that do O&M work on the system. These are made up of persons who cultivate

regardless of whether or not they own the land they till (Geertz, 1967).

[8]Such divisions need not lead to conflict, as observed in the Gal Oya scheme in Sri Lanka, where Sinhalese farmers are at the head of the system and Tamil farmers are located at the tail. The communal tensions and violence that have racked the country in recent years have not hindered cooperation within and between farmer groups. Some Sinhalese farmer-representatives have even gone out of their way to help and protect Tamil engineers and neighbors, and joint channel rehabilitation efforts have been undertaken despite violent incidents before the farmer organizations were started. One farmer-representative stated his own and his neighbors' view succinctly in a meeting in January 1982: "There are no Sinhalese farmers, and no Tamil farmers -- only farmers." (Uphoff, 1986).

[9]Solidarity is a very important factor in the operation of a water users association. In the Chhatis Mauja scheme in Nepal, the membership is quite diverse in terms of caste, yet it manages to maintain quite effective cooperation. One indication of the deliberate effort made to ensure that all members conduct themselves as equals and contribute equally to the work is the prohibition on anybody bringing an umbrella to maintenance activities. Some high caste persons might try to play the role of supervisor, as in earlier times, sitting under an umbrella and doing less work. This rule ensures that no one can act in a superior status role (P. Pradhan, 1984).

[10]Having a nominally "democratic" political ideology is no guarantee that government staff will be responsive to farmers' needs, or that they will not exploit farmers through corrupt practices (Wade, 1982a). The system of corruption that Wade (1984) documents includes politicians as well as administrators and technicians.

Part Three

Supporting Farmer Organization and Participation

7
Policy Considerations

Decisions about farmer organization and participation in irrigation management should not be made "across-the-board." This should be evident from our preceding discussions of:

-- the diversity of objectives in irrigation (Chapter 2)

-- to be met through various activities (Chapter 3)

-- at different levels of operation and organization (Chapter 4)

-- through persons having divergent characteristics who contribute in many kinds of roles (Chapter 5)

-- under heterogeneous conditions (Chapter 6).

Such analysis should help to clarify opportunities in planning and design for improving irrigation system performance.

No exposition can deal with all of the particularities and permutations for given situations. However, some generalizable directions can be formulated. Building on the foregoing analysis, it is possible to distill from the varied case study experiences some judgments about how best to proceed in improving irrigation management with farmer involvement. For this, we look in turn at:

(a) decisions about strategy and resource allocation which arise at the level of policy, setting the framework for developing water user capabilities to engage in management (Chapter 7)

(b) design alternatives and approaches which shape the organizational structures through which farmer participation can be strengthened (Chapter 8) and

(c) improving relations between technical staff and farmers as a requirement for bringing about more effective farmer organization and participation (Chapter 9).

The orientation toward farmer participation is not one of maximizing its extent, but rather of identifying and promoting kinds and degrees of participation which will further certain irrigation objectives. Such an approach is supported by the most quantified comparative study of irrigation system performance we found in the literature. Montgomery (1983) examined 20 irrigation case studies and judged them according to how well they handled three tasks: water allocation, land preparation, and fee assessment.[1] He found farmer participation improving the first activity much more than the second. However, for all three activities, the percentage of cases in the "good" category (compared with "fair" or "poor") was higher when decision-making involved elected or appointed bodies of farmers than when decisions were made solely by a bureaucratic entity.[2]

We are not suggesting here that participation by farmers will automatically improve irrigation performance -- or will improve it equally in all areas. As stated previously, no goal of 'maximization' of participation is warranted. But in most present situations, the amount of participation is so definitely 'sub-optimal' that it could be productively increased in a number of ways and at several levels.

Such a realization is gaining support in government agencies and among donors around the world. A review of World Bank experience found plans or policies to devolve major O&M responsibilities to water users projects in South Korea, Thailand, Niger, Senegal, Morocco, Brazil and Peru.[3] The outstanding issues revolve not around whether to accept greater farmer responsibility but rather up to what level (Chapter 4). There is considerable consensus that farmers should have rights as well as responsibilities "below" the outlet or turnout, at the field channel or water course level. How much higher in the system they should be able to make decisions, resolve conflicts, etc. is a matter of less widespread agreement. The Philippine National Irrigation Administration, which has pioneered participatory irrigation management, now accepts full farmer responsibility in "communal" systems and is experimenting with giving this over also to farmers cultivating 2,500 acre zones within "national" systems.[4]

Support

Political support from national leaders can be identified as a precondition, a sine qua non for increasing farmer participation in water management. Certainly political opposition from high levels, and even low levels, can stalemate or abort an effort in this direction. The best case of a national program shifting to an explicitly participatory approach is that of the Philippine National Irrigation Administration (Bagadion and F. Korten, 1985), where we find such support. But "support" is not a simple or a unitary attribute. The head of the National Irrigation Administration favored the approach, for a number of reasons including the practical need to recover from farmers the capital costs of NIA's upgrading "communal" schemes because of a change in the government's fiscal policy. But it was up to his Deputy Director and others to formulate and implement the agencies innovative approach. The necessary support was not a "lump-sum" or from any one individual. It involved willingness over a sustained period of time to acquire and provide financial help, to take initiatives in legislative and legal arenas, to procure and motivate adequate staff, and so forth.

Many calls for "support" are political equivalents of the "big bang" theory in astronomy regarding how the universe was created. But programs are seldom created all at once, even if they have a visible and dramatic "authorization" at some point in time. Conceptions and decisions evolve, and even once a program is launched, its continuation and further evolution depend on recurrent inputs and ideas. The cooperation of diverse persons and agencies is needed to move a program forward. A mechanism like the Communal Irrigation Committee set up by NIA to oversee and guide the introduction of participatory irrigation management can be useful for mobilizing and maintaining support from important sources.[5] Yet the form which such a mechanism should have, as well as its membership, will vary according to the circumstances and possibilities. "Support" should not be seen as some deus ex machina which empowers program initiatives, but rather as a sustained willingness on the part of key decision-makers in a program's environment to bear and share the political and economic costs entailed in achieving benefits from the new approach over time.

Support is needed from many different agencies and actors. They need to be kept informed about the program, its goals, its methods, its situation, its potential, its progress, its failings, and the obstacles it faces. The establishment of some kind of a support network of interested individuals, who will be able to affect the actions of key institutions is probably one of the most basic strategies for enlisting and maintaining cooperation with the program (D. Korten, 1982). Cooperation is what support must translate into; the latter is not a single thing, from a single source at a single point in time.

Support should be seen as something to be earned -- not a right but a result. Having high-level support is an advantage at any time in a program's development, but cooperation forthcoming because commands are issued from on high will not be as sustainable as that which arises from understanding and agreement at various levels. One often hears the lament that "if only" support were provided from the top, obstacles could be swept away. This neglects the lesson that practical demonstrations of accomplishments are more persuasive than words. While some high-level support is needed, cooperation is necessary at many levels. Progress which is consistent with the interests and values of key actors nurtures support, which is a vital concomitant rather than a prior condition.

In the area of irrigation management, the ideology of the regime does not appear to be a crucial determinant of support for farmer's participation. Some of the governments most favorably disposed toward a large role for farmers in irrigation management such as in Taiwan and South Korea are usually classified as politically conservative. No presumption should be made about whether a regime will support or oppose participatory water management. There are good reasons for any kind of government to favor this approach, especially if organizations steer clear of partisan politics, as discussed below.

Experimentation, Phasing and Flexibility

A "blueprint" approach to project design is no more appropriate than is the "big bang" conception of policy support for new programs. Experience in the Philippines, referred to above, and introduction of farmer organizations into the Minipe and Gal Oya irrigation schemes in Sri Lanka supports the contrasting "learning process" approach (D. Korten, 1980). Such a strategy of institutional development discards the conventional method of designing a program in advance and then implementing it as designed. Instead, it emphasizes problem identification and problem solving. Farmers play a positive role along with officials and engineers in such a process, since they understand problems and possible solutions even if they cannot articulate all the technical theories and details.

Programs for increasing farmer participation in irrigation management must be suited to the socio-technical context, which can vary considerably within a country (or even within a project area). For effective operation and spread, a program requires a cadre of persons who have knowledge, experience and commitment relevant to program goals, able and willing to adapt activities to circumstances. Even if the objectives are reasonably clear and widely agreed upon (the exception rather than the rule), the effectiveness of the means chosen to pursue them must be examined and tested empirically, to see how well they can work in a particular environment. In fact, both means and ends should be subject to continued assessment, in order to ensure that the efforts directed toward goals of social change are productive, and that the goals themselves remain appropriate. Such assessment requires involvement not only of top leadership but a capable and dedicated cadre of staff with open channels of communication.

This is not the place for an extended discussion of the theory and practice of "learning process." However, it should be said that this is not simply a matter of raw trial-and-error, since it involves some planning, even if tentative, and draws on bodies of experience and theory that appear relevant to the situation. It can be characterized as "inductive planning" (Esman and Uphoff, 1984:262-265). The approach emphasizes two things:

(a) proceeding according to <u>phases</u>, concentrating first on learning to be effective, then to be efficient, and finally to expand the scale of activity (D. Korten, 1980); and

(b) working <u>flexibly</u>, making modifications in light of experience and increasing understanding of the situation so that the prospects for resolving problems and achieving desired results can be increased (Rondinelli, 1983).

The policy implication of such an approach is that programs not be locked into rigid time frames or methods and that governments maintain a certain patience, forbearing from trying to "run" before an ability to "walk" has been demonstrated. The concept of "scaling up" is more relevant than "replicating," as the latter implies multiplication of organizations through a "cookie cutter" approach.

Building From Below

Prescribing a particular approach would not be in keeping with the preceding analysis and findings, but a general policy orientation does emerge from the case studies, seeking to develop farmer organizational capacity for irrigation management "from below." As suggested in Chapter 4, the terminology commonly used in describing levels of irrigation systems seems inverted. The highest level may be regarded as "primary" in terms of water allocation and division but not in terms of water management. Establishing water user organizations from the top down suits the administrator's perspective better than that of the user. The field channel level -- the unit command area -- is the common denominator in all irrigation systems. As pointed out in Chapter 4, it tends to encompass areas between 50 and 100 acres, although in intensively cultivated systems with small landholdings, the basic area of control and coordination may be 30 to 50 acres.

While larger units of organization should be able to aggregate greater amounts of labor and money, per capita resource mobilization is likely to be less. Such organizations are more susceptible to the negative effects of "free riding," as feelings of solidarity and mutual responsibility will be weakened by the attenuation of personal relationships when groups are larger. The costs of decision-making, communication, and conflict resolution become greater as the size of organizations increases. This means that developing greater capacity for planning and implementing water management -- not just water delivery and system maintenance but also cropping calendars, plant protection, etc. -- will be more feasible when the basic units are fairly small.

Farmer organization for improved irrigation management should be seen as a structure in which the groups at the unit command area level constitute the "building blocks." From these, a structure can be organized as high as is necessary. But one cannot expect the structure to be very strong if its basic units are weak or non-existent.

As suggested above, only some of the tasks of improved water management can be accomplished at the field channel level. To tackle the others, there is need for aggregating resources and capacities at higher levels. By linking base-level organizations vertically through some kind of federation which follows the physical structure of irrigation systems, it is possible to achieve "economies of scale" without undermining the field-channel-level units. This will create capacity for decision-making, resource mobilization, communication and conflict management both from above and from below.

One advantage of such an approach is that the lower levels of organization can be less formal and less legalistic, while higher levels can take on more of such characteristics to link better with government and financial institutions.[6] The lowest level of water user association can then more easily be absorbed into whatever are the residential, kinship or other patterns of consensus-building and conflict management in the community.

An alternative approach is to mandate organization from above by law, giving the effort to improve water management a formal-legal coloration. Because of the difficulty of setting up organizations everywhere quickly and uniformly at the lowest levels, formal-legal efforts are likely to be focused at the second or even third level above the field channel.[7] If created by legal enactments, groups are more likely to seek cooperation by threat of sanctions than by education and persuasion or by mobilizing community consensus and pressure. Moreover, in such situations, leaders are likely to be selected by government agencies rather than by farmers. As discussed in Chapter 8, there is reason to have some legal basis and powers for water user associations. The question is whether legal arrangements should be treated as a prerequisite for a farmer organization strategy or should serve as a supporting element. The latter view appears more tenable.

Existing Organizations

A major policy consideration concerning farmer participation in irrigation management is whether or not to work with organizations that already function among water users. There are situations where no such organizations exist, as in the Gal Oya project in Sri Lanka, but the issue often arises. Existing organizations may be "traditional" in their modes of operation. Some are specialized for dealing with water management, such as the subaks in Indonesia. Alternatively, water management activity may be associated with multi-functional organizations that operate as indigenous forms of local government, e.g., the council of elders of Daudzai in Pakistan, or the civil-religious authorities in Peru. Such local government bodies may oversee the work of persons in specialized roles handling water management duties such as the chawkidar at Daudzai or the ulu-ulu in Javanese villages in Indonesia. Or specialized water management groups like the tramos in Oaxaca state of Mexico may operate as part of the "traditional" local government structure.

The policy choice pointed out here may intersect with decisions about organizational design, discussed in the next chapter i.e., whether water management should be handled by specialized or multi-purpose organizations. But the choice is a separable one, with substantial implications for the kind and pace of organizational development. As a rule, the preferable course is to work with and build upon existing organizations at the local level, subject to the qualification that equity objectives are not seriously compromised.[8]

Existing organizations are not necessarily "traditional." A review of water user associations in World Bank projects found that the WUAs co-opted into a project in Afghanistan were based on roles going back many centuries. But in the Moroccan and Peruvian cases considered, the extant associations were operating under modern legal procedures. In all these cases, the WUAs could make contributions to design, construction, operation and maintenance activities with little "start-up" investment (Cernea, 1984).

There has been a tendency for donors and governments to ignore existing social organization in project planning. The Palsiguan project in the Philippines is a clear example of this, where approximately 20,000 of the proposed 25,000 acres to be given more integrated and assured water supply were already cultivated and managed by more than 170 user groups (_zanjeras_) with an impressive and extended performance record (Coward, 1985a). The donor-assisted plans which redesigned the water distribution system for the area without regard to the existing organizational arrangements amounted to "disinvestment" in social capital represented by the indigenous WUAs. Fortunately, this approach which ignored the _zanjeras_ was reconsidered by the agency (Visaya, 1982).[9]

"Traditional" organizations in the modern world are seldom encapsulated. A study of how irrigation is managed in the Taita Hills of Kenya suggests that "indigenous modes of organization (the patrilineage and the neighborhood) ...merge almost imperceptibly with modern organizations (the sub-location and location)." Canal committees, special-purpose irrigation organizations which operate informally, are chaired by lineage elders. Disputes that cannot be resolved within the jurisdiction of a canal committee are taken to sub-chiefs or chiefs who are legally-recognized authorities within the local government structure (Fleuret 1985:114).

As noted in Chapter 5, indigenous organizations handling irrigation in some countries of Latin America have been undercut by government policy. Whether their capacity could be restored by more favorable official attitudes now is not certain. But the tradition of local management can find expression in non-traditional institutions as seen in rural Peru and Bolivia.[10] What is most valuable about existing organizations is that they already have procedures for decision-making, patterns of communication, and means for building consensus and resolving conflicts. These are capacities that invariably take some time to develop anew under the best of conditions.

Ownership

In undertaking participatory irrigation management, the question "whose system it is" needs to be addressed. To the extent that water users feel some proprietary interest in the system, they will also be more responsible in their use and preservation of it. Ownership is a very complex matter, implying responsibilities as well as rights. Property rights themselves range from simple use (usufruct) to the right to sell, transfer by inheritance or even destroy the property in question. Often certain obligations are attached to rights, such as the requirement that owners of land getting water from an irrigation system contribute labor and/or materials to its upkeep.[11]

Property rights can be established in various ways -- by precedent of use, by investment of one's own resources in creating the asset, by purchase from someone who claims the asset, by decree of the state. Where users are owners of the water as well as of the land, they derive greater benefits from the process of production. This gives them both more incentive and more responsibility to maintain the system properly, to operate it efficiently and even to extend it.

This is seen when comparing farmer performance in "communal" irrigation schemes and in state-owned ones. In the cases we reviewed, users invariably accepted and carried out greater responsibilities for all aspects of water management in the first type. In light of this, it is surprising and disappointing that the trend in developing countries is for expanded state control and claims of ownership in irrigation systems (Coward, 1983 and 1985). In Indonesia, for instance, when government agencies moved into community-operated schemes to "modernize" them, they took over responsibilities for O&M previously discharged by farmers (Morfit, 1983). This is now being reversed on an experimental basis, following the example of the National Irrigation Administration in the Philippines which has pioneered in preserving and even expanding water user ownership and responsibility.

The three-fold distinction introduced in Chapter 3 is important here, as "ownership" may refer to the <u>water</u> in question, to the facilities that acquire and distribute it, or to the organizations that manage the water. Rights of ownership to water are conditioned by the fact that no one can claim to have created it. Some public interest in the allocation and use of water may be asserted by the state with normative as well as practical justification.[12] Where water users have created the facilities that capture and convey water, their claim of ownership is strong. The question arises whether irrigation structures constructed by the state, with public funds, should be treated as state property or as public property, with users regarded as members of the public. The latter view may be accepted as a matter of policy for the sake of better irrigation management.

Water user associations of any kind, even if established at the initiative of the state, should "belong" to their members and not be regarded as state institutions. Government personnel can, with enough expenditure of funds, operate and maintain physical facilities, even down to the turnout and below. But they are much less able in any case to carry out at all levels the organizational tasks we have identified for water management. Just as farmers feel more responsible for the O&M of physical facilities if they consider themselves the "owners" of the structures, so they will invest more time and effort in organizational activities like communication and conflict resolution if such responsibilities are fully "theirs" and if the organization can be used to further their interests.

Non-Political Associations

A very delicate but important subject is whether water user associations will be kept divorced from "politics," especially partisan politics where these operate in the countryside. Water is a valuable and usually scarce resource. Where privileged access can be gained to it through political manipulations or influence, there is great temptation to do so. Yet almost universally we find that "politicization" of water allocation and distribution leads to irreconcilable conflicts, reducing the effectiveness of water user associations and often causing their demise.

Water users generally recognize this. In Nepal, it is reported that members of the management committee for the Chhatis Mauja scheme cannot, while in office, have any active role in local politics, and farmers in the Magar Kulo scheme said that that any external interference from the district or village panchayats in their irrigation activities "was not to be tolerated" (U. Pradhan, 1982:5). Farmers in the Gal Oya system in Sri Lanka have tried to keep their personal political attachments out of organizational matters.[13] Peasants in the Bakel irrigation perimeters in Senegal have emphasized that their farmers associations are "non-political" organizations (A. Adams, 1977).

In some of the cases reviewed, e.g., in Tamil Nadu and Andhra Pradesh states of India, water user associations did make contact with politicians such as members of parliament to promote their interests with higher levels of government. But they were more effective to the extent that their organizations were not internally divided along political lines, so groups could not be played off one against the other. Within the Muda scheme in Malaysia, it was assumed by the government that all associations would remain within the majority party framework, in which case "politics" did not arise. The same is true for the irrigation associations in Taiwan, where representatives of the governing party act as "watchdogs" to keep the WUAs reasonably honest and effective, guarding against the kind of malfeasance by officials which contributed to its downfall in the mainland.

Where there is a dominant party, the decoupling of politics from water management is itself "good politics" if it improves the efficiency of irrigation and enhances citizens' satisfaction with their government. Keeping partisan politics out of water management is more difficult where party competition is keen. Yet as seen in Sri Lanka, a country with strong traditions of partisanship, farmers may themselves strive to remain independent, to distribute water according to individual rights and needs, knowing that once partisan considerations intrude, the cooperation among farmers which is so necessary for proper irrigation will crumble.

No matter what farmers want, if the government seeks to politicize WUAs in a single-party-dominant or in a competitive party system, it can easily do so by injecting partisan criteria in water distribution. WUA members can have a "truce" among themselves with regard to furthering their respective party loyalties. Sri Lanka has had fierce party competition for over 30 years, yet Gal Oya farmers have explicitly barred (or balanced) party considerations in the operation of their organizations. The policy question for governments is whether they will refrain from seeking narrow partisan advantage, as the Sri Lankan government has been willing to do thus far, with positive results for itself and for farmers.

Bureaucratic Reorientation

One of the main deterrents to farmers assuming or accepting more responsibility for irrigation management is the orientation of the agency personnel with whom they must work. We find many apparently justified complaints by farmers of non-performance or malfeasance on the part of irrigation bureaucracies in settings as diverse as Brazil (Hall, 1978), Egypt (CSU/CID, 1980) and India (Wade, 1982a). If the staff dealing with farmers are arrogant, indifferent or corrupt, water users will be hesitant to work in a collaborative mode with the personnel of government agencies. One should not criticize or indict all irrigation engineers, technicians and administrators -- a great many are cooperative, hard-working and scrupulous. But too many, whatever their competence, appear unwilling to work respectfully, seriously and fairly with water users in a joint enterprise of improving irrigation efficiency.

This problem has given rise to the concept of "bureaucratic reorientation" (BRO), reflecting experience in the Philippines and Sri Lanka with changing irrigation bureaucracies' attitudes and practices toward farmers (D. Korten and Uphoff, 1981). Bureaucratic reorientation as discussed further in Chapter 9 is not achieved by orders or indoctrination. To create a more positive attitude toward participation, the approach itself should be implemented in a participatory way, involving engineers and technicians in a process of collaboration with farmers, political leaders, administrators and consultants whereby the agency's doctrine and self-image are modified to support a new relationship with water users. For example, the agency's "mission" may be revised from one of assuming complete responsibility for all water allocation and use down to the farm level to one of distributing agreed volumes of water to intermediate levels of the system, from where user groups assume responsibility for its distribution. Such a shift of self-styled responsibility from "retailing" water to "wholesaling" it implies considerable relaxation in an agency's attitude with respect to command and control, refocusing efforts toward reliable distribution within the main system.[14]

Aside from engendering mutual respect among farmers and technical personnel and a new sharing of management responsibilities, BRO seeks to draw on the technical contributions of farmers. We find an increasing number of documented cases where farmers' advice to engineers who were designing and building dams or canals was ignored, with unfortunate consequences. In a Mexican case, for example, technicians were building a dam based on rainfall records for only the last 15 years. Farmers knew that flash floods might destroy it in that location, but their knowledge was not sought. Soon after construction, just such a mishap occurred (Cernea, 1984a).[15] Similar cases have been documented in the Philippines and in Nepal (D. Korten, 1980; Shrestha, 1980).

Where there is "dialogue" with farmers it is too often perfunctory and ineffective. Coward (1985) reports from Indonesia where government staff have been rehabilitating irrigation systems and putting in new division boxes, many of which do not function properly:

> The problems include the following: (1) boxes incorrectly located, (2) not all the canals from the boxes have been constructed, and (3) the design of the openings does not allow accurate delivery as required by the (distributional) rules. This is not surprising given the design process that was used.[16]

The technicians went through the motions of consulting the irrigators in these systems before rehabilitation work began, but obviously they invested too little time, attentiveness and skill in the "dialogue" to make it worthy of the name.

Bureaucratic reorientation, whereby government personnel become more willing and able to interact constructively with members of the public, particularly with organized groups having development responsibilities, will not happen without policy support. As indicated earlier in this chapter, sustained and diffuse support is needed, not just some advance pronouncement. BRO should not be viewed as a precondition for establishing farmer participation and organizations for better irrigation management. It does not occur all at once or in a vacuum. It is an evolutionary process, in which demonstrations of farmer competence and conscientiousness can encourage changes in officials' attitudes and performance -- and conversely, changes from the government side can encourage farmers to assume and discharge greater responsibility. This process is so important that it is considered more fully in the concluding chapter.

Compatibility of Objectives

Probably the overriding aspect of policy affecting farmer participation is the extent to which the government's and water users' goals coincide. We have noted in Chapter 2 that there is often divergence, and that water users themselves can have competing objectives. Conflicting aims are likely to impede cooperation among farmers and with government agencies (Svendsen, 1983). A government can legitimately seek higher production with a view to having more food to feed the whole population, or to increasing exportable surplus to earn foreign exchange, while farmers may seek to maximize their returns by investing labor in activities other than irrigation (e.g., not keeping channels cleaned to a technical standard).

When the objectives of the government and of its citizens come into conflict, one cannot say automatically which should take precedence. Regimes can be short-sighted or pursue narrow interests just as surely as may sectors of the public. Each conflict needs to be examined in light of the competing justifications offered, and any reconciliation between them is more likely to reflect the balance of respective power between government and various sectors than some public ordering of values. Without prejudging outcomes, we note in concluding this chapter that a crucial policy consideration is the extent to which, in any irrigation management situation, the government wants for water users what they want for themselves, their families and their communities.

If governments expect farmers to take responsibility for maintaining facilities which do not "belong" to them, or seek to make water user associations serve partisan purposes, for example, various economic, technical and social objectives of improved irrigation management are likely to be undermined. We do not expect that regime and user goals will always be congruent, or need always to be the same. Conflict is a common condition of social existence. **Channels of organization, among water users and between them and government agencies, have as one of their main purposes the identification and resolution of competing objectives, to build a basis for cooperation despite divergence of initial positions.**

Irrigated -- in contrast to rainfed -- agriculture establishes greater incentives for reconciliation of contending interests. Not only are farmers more interdependent, but officials' ability to achieve the results they expect from investment in irrigation will depend crucially on the efficient performance of water users. As this requires a considerable degree -- and under present conditions, usually a greater degree -- of farmer participation and organization, we review some design considerations for promoting this in the next chapter.

NOTES

[1]Half of the cases were large-scale projects in India for which he could find documentary data; the rest came from Indonesia, Kenya, Mali, Mexico, Pakistan, Spain, Sudan, Taiwan, Thailand and Zimbabwe. Six of his cases are included in our set of 50. The activities assessed under "land preparation" are not clearly stated -- Montgomery describes it as "a maintenance and operations function." Fee assessment involved conflict management as well as resource mobilization.

[2]Montgomery's hypothesis was that participation would produce the best results when an activity required: (1) much adaptation to variations in local conditions, (2) decisions made frequently but not routinely, (3) quick response to unanticipated events, and (4) substantial changes in behavior. According to these criteria, he expected water allocation activities to benefit most from user involvement, and they appear to do so. But some advantages from farmer participation were evident even for land preparation and a substantial gain was seen in fee assessment. The percentage of cases in the "good" category for the three respective kinds of irrigation activity was 75, 25 and 67 with farmer participation, compared with 58, zero and 13 when bureaucratically managed. The percentage of cases judged "poor" was zero, 38 and zero when farmer representatives were involved, compared to 58, 75 and 56 when bureaucratic entities had responsibility.

[3]See Section III of Cernea (1984). SAED, one of the irrigation agencies in Senegal, for example, plans that "in the longer term, farmers' cooperatives will replace SAED in perimeter (village irrigation scheme) management." Farmers' groups in South Korea are the most advanced. Cernea reports that they "were adept enough at water management skills to achieve equitable water distribution among participating farmers even when all construction works had not been completed at the same time."

[4]Personal communication from Benjamin Bagadion, recently retired Deputy Administrator for Operations of NIA, in seminar at Cornell University, June 1, 1985. The legal authorization for this has been "on the books" for some time, awaiting the build-up of water user associations, farmers' competence, and engineers' confidence in such an approach.

[5]On the CIC, see D. Korten (1982), F. Korten (1982), and Bagadion and F. Korten (1985).

[6]One of the findings of a quantified comparative study of rural local organizations was that informal modes of operation were more often associated with effective performance than highly formal modes (Esman and Uphoff, 1984:141-144). While larger organizations were slightly, but not significantly, correlated with better performance (larger size was probably as much a result of success as a cause thereof), good performance was strikingly and significantly associated with smaller organizations when these were linked vertically with higher levels of organization (pp. 149-151).

[7]This is discussed in Uphoff (1982). Such an approach in the case of Pakistan is discussed in Radosevich (1975) and Cernea (1984). Promising results are reported there by Fairchild (1985), but he acknowledges that less than half of the 10,000 WUAs thus created may be functioning adequately.

[8]This is essentially the position taken in the USAID policy paper (1984) on the contribution of local organizations in development. It is also the conclusion of an analysis for the World Bank by Kottak (1985:342) from looking at Bank-assisted irrigation projects.

[9]In Thailand, it is reported that when the government undertook to "modernize" existing irrigation systems by upgrading the physical structures it sought to impose a uniform system of local irrigators' associations, with standard by-laws, centrally sanctioned personnel, etc. However, farmers had already developed and operated the existing irrigation systems for generations through their own local associations,

> Each (of which) has its own approach to selection of leaders, organization of maintenance tasks, regulation of water use, punishment of cheaters, flood protection, etc. If notions of bureaucratic 'efficiency' dictate that all such associations be reorganized to fit some externally designed template, there is grave danger that local skills will be blunted and irrigation water used less efficiently... Any decision that imposes ex post local uniformity (e.g. by threatening to cut off resources) should be supported by evidence that local efficiency and production will be enhanced, and not by unspoken, aesthetic judgments regarding bureaucratic order. (Calavan, 1984:221-222)

[10]"In the four communities I visited (in Peru), traditional governing and irrigation institutions had been replaced by

campesino community structures. Despite this change, local institutions in three cases did a more than adequate job of mobilizing community labor. Community presidents acted much like small town mayors in the United States, soliciting outside development funds in return for offers of a hard-working, mobilized labor force." (Lynch, 1983:10) In Bolivia, CARE was developing small-scale systems for irrigation and potable water with the help of Clubes de Madres (mothers' clubs), having a formal agreement to work with them where they exist and to get financial support from them. Where no Club exists, CARE works with community cooperatives (1983:15).

[11]The zanjera organizations for irrigation in the northern Philippines, for example, when constructing headworks and channels to serve new command areas vest rights to shares of water (called atars) in those who have contributed resources to the work. But continued receipt of water is conditional upon continuing to contribute labor and materials for O&M (Siy, 1982). Similar requirements are found in most other communal systems (e.g., Gray, 1983; Mitchell, 1976; Martin and Yoder, 1983).

[12]Irrigation is an activity which benefits and affects more than just the irrigators, so some state involvement in water rights can be justified. Downstream users, for example, may need government support in preserving their access to water. This however creates a basis more for regulating use rather than for taking over property rights.

[13]Several politically influential farmers who initially opposed farmer groups in Gal Oya, apparently for fear of losing "clout," came to accept independent, non-partisan organizations there. One leader, previously chairman of the village council, president of the multi-purpose cooperative, and campaign manager for the member of parliament, said he opposed any political links for the water user associations, because "Politics is cancer for water management" (personal communication, June 1982). When Gal Oya farmers organized a "convention" in October 1984 they invited the Ministers of Lands and of Agriculture, as well as the District Minister to come as guests. But they decorated the platform in a politically neutral color (white) to indicate their non-alignment in party terms.

[14]Agency efforts to "retail" water are seldom very effective anyway because of the knowledge and personnel requirements for distributing water down to the lowest level. Deficiencies in main system management are generally more significant sources of water loss and inefficiency than is users' water management at

lower levels of the system (Wade and Chambers, 1980). In the only quantified analysis of the comparative efficiency of water distribution at different levels of a system, using detailed water and yield data from four seasons in Gal Oya, Sri Lanka, Wijayaratna (1986) has found that proper "wholesaling" of water (optimizing distribution within the main system) could raise production within the system, as it was operated in the early 1980s, on average by 13 percent. Since such gains would be a pure windfall, requiring no additional inputs by farmers, the average net return (income) would be thereby increased by 37 percent. Improvement in farmers' "retailing" of water -- not yet optimum in the early 1980s because WUAs were only just being introduced -- could of course add to these gains.

[15]"The floodwaters swept away many irrigation wells and several fruit tree plantations, some of which were PIDER [project] investments. The experts who had been responsible for the dam's construction explained to the central [government] inquiry team that its collapse was an act of fate; the planners had consulted the hydrometric series for rainfall and surface run-off for the previous 15 years. The inquiry team, however, also consulted the villagers. Their answer was that the experts had not paid attention to local experience. The oldest villagers recalled clearly that 30 years earlier it had rained for two complete days, and the stream had risen to a level two meters higher than the maximum calculated by the experts." (Cernea, 1984a:41)

[16]Coward (1985) adds: "An external team in the field for a short time and with little conversation with farmers ... was led to a new physical apparatus that simply does not fit the principles of water distribution found in the original system. It is unlikely that the design team was even aware of these principles.... In [one] case, a new division box divided a canal flow but in a ratio unlike the original. To correct this problem, the water users took the two stream flows from the new box, joined them together again, and then placed an [indigenous division box] in that canal to correctly apportion the flow."

8
Choices in Organizational Design

While water user associations represent a major means of improving irrigation management by expanding farmer participation and responsibility, they are not always or everywhere effective. World Bank irrigation projects increasingly provide for a WUA role (Cernea, 1984), and USAID's policy (1984) supports such organizations. Where WUAs are not helpful, there are usually good reasons why not, stemming from the way they were conceived, structured and introduced by non-users. This chapter focuses on what kinds of WUAs to assist, and in what way. The conclusion in the preceeding chapter about working with existing organizations applies. "Strengthening" where possible is more desirable than "introducing." Still, the latter may be the only course available. The following discussion of design choices is presented in terms of new WUAs, but as we are dealing with principles rather than specifications, the propositions should with appropriate modifications be applicable to working with existing organizations as well.

Understanding irrigation systems in socio-technical terms suggests that organizational design efforts should not be entirely separate from technical considerations. Inappropriate technical decisions and performance can certainly make futile even the best of organizational intentions and plans.[1] Equally important, organizations cannot operate without satisfactory material possibilities and rewards. Water user associations must have appropriate capacities to accomplish specific tasks if they are to be productive and sustainable. One should not expect WUAs to function in the same way or to have the same arrangements as a parent-teacher association, for example.

Choices about organizational design should be made in collaboration with technical personnel, even if planning and implementation efforts are the responsibility of specialists who deal with social relations and with developing institutional capacity for irrigation tasks. As discussed in the next chapter, we would not suggest any simple division of labor between social scientists and technical personnel, with "participation" delegated to the former and all engineering and other decisions handed over to the latter. Lead roles need to be assigned, but what is sought is a meshing of organizational and technical provisions. The following discussion focuses on the former, but it presumes that there is consultation with engineers and administrators -- as well as with farmers themselves.

One major design issue is whether greater farmer participation can be promoted without introducing some change in the physical system. That is, can there be sufficient incentives even without construction or rehabilitation of structures and facilities (with water users involved in processes of design and construction) for farmers to assume more O&M responsibilities? There is not enough systematic evidence to answer this question conclusively, but practically all of the government-supported initiatives to expand water user roles in irrigation have been accompanied by some investment in physical changes. It is reasonable to expect that patterns of behavior are more amenable to change when physical relationships and possibilities are in flux. The fact that a government is investing in improving system performance could provide an incentive for water users to be willing to contribute to such an improvement.[2]

Scope of Activity

Whether water user associations should be limited to irrigation management activities is a first-order design issue. Since projects are usually administered by an irrigation department, it commonly prefers such a restriction if only to avoid "encroaching" on other agencies' bureaucratic turf. But from farmers' point of view, such boundaries are likely to be irrelevant. Obtaining the full benefits of better irrigation practices requires improvements in extension, input supply, even roads; and organized farmer participation may be important for these as well as for better water management. On the other hand, the latter is itself an immensely complicated undertaking as analyzed in Chapter 3, and non-irrigation tasks can overload an organization's capacity.[3]

Water user groups are likely to be multi-functional to some extent because of the complexity of irrigation. Indeed, denominating the number of "functions" performed by an association is itself a difficult matter. Other things being equal, the quality of performance of local organizations appears to be somewhat greater for multi-functional organizations (Esman and Uphoff, 1984:139-141). This could, however, be a matter of more effective organizations, with capable leadership and strong membership support, taking on more functions. One should not infer that giving organizations more tasks will in itself make them more successful, even if there is a correlation between better performance and more tasks being performed.

That water user associations are such a common type of specialized rural local organization suggests the validity of their having limited scope. The fact that irrigation lends itself to the "membership" form of local organization (P. Doan et al., 1984: 7-10, 21-23) also suggests that groups should focus on water management activities. Of course, if lower-level irrigation management is handled through local government institutions, an option discussed in Chapter 5, the local organization responsible for water management will definitely be multi-functional.

In his review of WUA experience, Cernea (1984) finds organizations handling a wide range of activities. The indigenous WUAs incorporated in the World Bank's Khanabad Irrigation Project in Afghanistan dealt fully with operation and maintenance for water allocation, distribution and drainage, and additionally took care of farm roads within the system. The traditional village government bodies brought into the Periyar Vaigai Irrigation Project in India exercised a similar range of duties, while existing WUAs in a Moroccan project did not go beyond water distribution and maintenance tasks. South Korean farmers' irrigation organizations, on the other hand, in addition to handling a full range of water management activities, are involved with extension and training and even group farming. Projects planned in Brazil and Niger where new WUAs were to be formed expected to undertake marketing, credit and other tasks. There are some good examples of multi-functional WUAs, such as those in Thailand cooperating with the "training and visit" system of agricultural extension promoted by the World Bank, where the WUAs provided the "contact farmers" necessary for that system's performance. Nevertheless, Cernea concludes that:

> organizing irrigating farmers for the single purpose of managing water resources (itself no easy task...) is a more recommendable strategy for water-based rural development than broad multi-purpose organizations, unless farmers themselves have had experience in combining their irrigation tasks with others such as credit seeking and sharing.[4]

In the Gal Oya case in Sri Lanka, extension staff found that the farmer-representatives chosen at the field channel level could and would function in the "contact farmer" role (Uphoff, 1985).[5] The decision on whether or not a specific WUA would engage in extension activities was left to its members, however. This seems to be the most appropriate way to resolve the design issue of "scope." The range and content of WUA responsibilities should be determined by the members themselves, which is only fitting if the organizations are in fact "theirs" as suggested in the preceding chapter.

The basic responsibility of WUAs should be water management, with all of its variety and complexity of tasks. Beyond this, members should be able and even encouraged to diversify into complementary tasks, in directions and at a pace on which they can agree. Indeed, as the tasks of water management come to be effectively performed and routinized, there is incentive as well as opportunity for WUAs to become involved with extension, credit, infrastructure and other activities. With good water supply, these other activities develop higher payoff.

Project designers should provide for WUAs an evolutionary path of institutional development. There should be no strait-jacketing organizational blueprints which have WUAs all doing the same things, since not all will have the same needs and capabilities. Minimum expectations can be stated, even required, but only to ensure some basic performance level. For WUAs to move beyond this and into non-irrigation responsibilities may require bureaucratic negotiating and trouble-shooting so that agencies outside the irrigation sector are willing to work with farmer groups in new ways.

Size and Structure

The debate over how large or how small WUAs should be is recast by the analysis presented in Chapter 4 and by the conclusion that WUAs at the field channel level should not be expected to operate in isolation but should be federated in some multi-tiered organization. The size of the lowest level of operation (water control) should usually be the size of the base-level group of water users. This area usually encompasses between 50 and 100 acres. When the lowest level unit command area is in the upper end of this range or exceeds it, one usually finds informal subdivisions, perhaps with temporary control structures establishing smaller groups of farmers who operate distinctly from others within the area.

In the Pochampad system in Andhra Pradesh, India, the Pipe Committees set up by the project authority covered about 100 acres each. They performed better once "zones" had been established within each area for purposes of water rotation and for sub-group representation in the committee (Singh, 1984). The Nong Wai irrigation project in Thailand, given its hydrological layout, had basic command areas (chaks) ranging between 100 and 1250 acres. In the larger chaks, farmers set up water user groups along respective ditches, so that 6-10 farmers could carry out rotations within a block of 30-50 acres. Each group received water one day per week and shared it among themselves. Each was represented in the Nong Wai Agricultural Cooperative Society which was established to integrate irrigation and agricultural development work.[6] These are examples of two-tiered structures of organization, but cases are reported in Chapter 4 of three, four and even five levels of organization being built up from small base-level groups.

In short, the size as well as the structure of water user associations should correspond to the hydrological features of the irrigation system. The most promising design will begin with small base-level groups, possibly operating fairly informally, whose size and area are set by the lowest level of water control. This level is determined organizationally as well as technically, since on a long field channel a group could build and operate a cross-regulator to establish a distinct command area for itself. We have found that base-level command areas tend to be around 50 acres, cultivated by 15-20 farmers.[7] But specific numbers will be affected by canal layout, topography, holding size, soil quality (difficulties of water retention), etc. Whatever size group under the circumstances can achieve the most efficient cooperation, communication, conflict management, etc. is to be preferred. As this will not be uniform within a system, let alone a country, WUA design should allow for variability of size.

The number of levels of organization will similarly be established according to the hydrology of the specific irrigation system. The principle of direct participation at the lowest level gives way at higher levels to procedures for indirect representation through spokesmen chosen by farmers to speak and act on their behalf.

Membership and Decision-Making

The term "water users association" suggests that membership is composed of those persons who benefit from a common source of water. As used by governments and donor agencies, it refers to irrigators. Yet in many instances, there are persons who use water from a reservoir, pump or canal who are not cultivators, which calls for a major design decision whether to try to include such persons in WUAs. Where agricultural and non-agricultural interests are compatible, this may present no difficulties; however when they are conflicting, WUAs may be unable to withstand the tension. In such a situation, the problem should be addressed as a matter of policy, to establish legal or other limits on the competing demands -- not as a matter of WUA design.

The main issue treated in the literature is whether WUAs should be composed of people who live together or who work in adjacent irrigated areas. Coward suggests that "for purposes of irrigation organization the critical unit is the irrigation community composed of field neighbors and not the village community composed of residential neighbors; though in some instances the two groups may be one and the same" (1979:5). Where the two sets of persons coincide, it should be easier to operate effective organizations, but where the two diverge, membership is more appropriately shared by neighboring cultivators. Of course, farmers may have several different groups of field neighbors when their holdings are dispersed.[8]

The cases where residential neighbors constitute the irrigation community are in situations where virtually all households depend on irrigation, and the local government can carry out water management functions in the name of the whole community, as in many Indonesian communities described by Geertz (1967) and Duewel (1982) or in the Indian villages studied by Wade (1979, 1982, 1984). Where water user organizations are village-based, as in Indonesia, it is reported that water management is better when villages cover entire (but small) hydrological units (Hutupea et al., 1979).

One particular problem concerning "membership," raised in Chapter 5, is how to ensure appropriate representation of disadvantaged cultivators, particularly those who own no land, since membership in WUAs is usually defined in terms of "household heads" or land ownership. Women tend to be consistently underrepresented in WUAs, as reported from studies in Africa, Chile, Peru and India (Lynch, 1985:43-45). This is true also in the Philippines where WUA membership is assigned to "household heads" (F. Korten, 1982:15-16). If membership were vested in "households" instead of in their "heads," it would be possible to have greater women's involvement.[9] Project designers should consider this as a means of expanding participation and making outcomes more equitable.

Ensuring a role for landless cultivators may be even more difficult because landowners may have reasons to want to exclude them. Formal WUA membership might give them a claim to secure tenancy status or even to land ownership under land reform legislation, for example. If the lowest level WUAs operate informally, not according to legally-defined membership and rules, it may be feasible to involve tenants and sharecroppers de facto if not de jure. Their participation in decision-making and operations is crucial when they rather than the landowners are the actual users of the water. This is not a fully satisfactory resolution of the issue but it should produce better results than assigning "membership" only to owners when designing WUAs.

When it comes to making permanent investments in the irrigation system, landowners should bear the full material and organizational costs since they benefit most from the resulting increase in land value. This might be handled by having two categories of membership. In some Indonesian systems, WUAs make differential work assignments according to land tenure status when undertaking short-term compared to long-term improvements (see footnote 15). As discussed previously, where there is great heterogeneity in tenure status, having effective WUAs is likely to be more difficult, though it is not impossible.[10]

The structure of decision-making and the way members become involved in it is an important design feature. The case studies tend to confirm the general finding from a broader analysis of rural local organizations, that the best structure is one with an assembly of all members who meet periodically, supplemented by one or more committees, possibly an executive committee, which can exercise more direct and active leadership (Esman and Uphoff, 1984:144-146).[11] The advantage of providing all irrigators with an opportunity to meet on a periodic basis and discuss problems, plans and policies is that information can be more readily shared and accountability of leaders can be better maintained. It is difficult for large bodies to exercise responsibility, however. Thus, having committees in addition to an assembly can combine the strengths -- and compensate for the weaknesses -- of each mode of decision-making. Such a structure has served well the Spanish irrigators in Valencia for over 700 years (Maass and Anderson, 1978). In the cases analyzed in Annex 3, about two-thirds of the organizations have some combination of committees of officers and an intermittent assembly. The one-fifth which had neither committees nor assemblies were far from the best examples of irrigation management.

Leadership and Responsibilities

Effective water user associations invariably require capable, committed and accountable local leadership, as discussed in Chapter 5. Project designers cannot ensure capability and commitment, but they can try to build accountability into the WUA system, for example, through the kind of decision-making structure established. Just as planners must consider whether they can and should work with existing organizations, they need to weigh the advantages and disadvantages of having existing community leadership fully involved in WUAs. There may be no choice, where such leaders are powerful in social and economic terms and well-connected politically. In such situations, efforts should be concentrated on having organizational checks and balances that provide for accountability, and on expanding the pool of prospective leaders by minimizing the tendency for a few persons to monopolize all positions, activities and experience.

It is desirable to mobilize new, agriculturally-oriented and civic-minded leadership wherever possible. This spreads responsibility, where present local leaders are likely to be overextended already. It also leads to some specialization in carrying out water management tasks, which can develop useful expertise within the community.

The methods for attracting and supporting a new cadre of leadership are still in experimental stages. Experience in the Philippines, Sri Lanka, Nepal, Thailand and other countries suggests the value of using organizers as "catalysts," as examined in Chapter 9, to enhance the prospects for mobilizing such leadership. This is consistent with the policy orientation of "building from below," discussed in the preceding chapter. It is very important to have conscientious and respected leaders in a WUA program from the very outset, since once a program is dominated by leaders who have less than laudable reputations, it will likely be difficult to attract more suitable individuals to come forward.[12]

The responsibilities of WUA leaders will vary according to the <u>level</u> at which they are working. They may operate either directly at the lowest levels of the system as agents of their irrigator-neighbors or indirectly as representatives dealing with decision-making and conflict resolution at higher levels. The design of farmer-representative roles should reflect the differences in activities and levels analyzed in Chapters 3 and 4.

There should also be realistic expectations about activities which match the pattern of seasonal variations. Uniform, year-round responsibilities, such as bi-weekly meetings, are alien to the cycle of the seasons. It should not be regarded as dereliction of duty for organizational work to slacken when competing labor demands increase for field preparation, planting or harvesting or between seasons when attention turns to non-irrigation activity.

Whether WUA leaders should be paid or not, and if so by whom, has been debated. Quite a variety of arrangements are reported in Annex 3. If leaders are paid it seems important that the salary or in-kind remuneration come from WUA members to secure greater accountability of leaders (Meinzen-Dick, 1984). Unremunerated leadership has been provided for in a number of cases, such as in the Nong Wai scheme in Thailand (Kathpalia, 1984). If the level of group solidarity is high and the WUA program produces some tangible results, having leaders work on a volunteer basis appears feasible.

Leaders' formal work requirements should be kept within reasonable limits. A World Bank project in Pakistan proposed and got enacted a law that enumerated WUA leaders' responsibilities. It detailed nine functions for each member of a WUA Board of Directors, which could amount to a full-time job. At the same time it prohibited Board members from receiving benefits greater than any other WUA member by virtue of their elected positions. Sanctions could be imposed by the project's managers if the functions were not properly executed.[13] Such an approach, while giving the appearance of control over water management performance, is likely to be counter-productive.

Legal Basis

The question of how to proceed with legal enactments to establish water user associations is an important one. There is reason to have some legal framework within which WUAs operate. Otherwise, lacking official recognition and standing, they can be ignored by administrators and engineers. Moreover, they may be unable to operate bank accounts or to raise funds for their activities. But as suggested in the preceding chapter, there is some question whether legal measures should be used to "create" WUAs or to undergird them after they have gained some initial momentum.

There are several dangers in a legalistic approach to setting up WUAs. If their existence is mandated, officials are likely to set up "paper" WUAs that satisfy the letter of the law. Purely nominal WUAs add little to water management capacity. Second, legal provisions tend to be uniform, i.e., all WUAs must have the same set of officers, the same size executive committee, the same dues, the same meeting dates, etc. Standard structures and procedures will produce WUAs that are sub-optimal because their terms of operation are not appropriate to each particular setting and because members will have a lukewarm commitment to what is not their own creation. Third, legal instruments often have set forth an asymmetry between rights and duties, stressing obligations of WUA members, in a way that dulls incentives for active participation.[14]

The successful program of introducing participatory water management in the Philippines began with some changes in the relevant laws, empowering groups of water users to take over responsibilities otherwise vested in the National Irrigation Administration. However, implementation proceeded gradually and experimentally. Once WUAs have been well established with the help of NIA, real "powers" are handed over to them, balanced by "obligations" that are reasonably well understood and accepted by members.

When legal frameworks are drafted, it is important that they prescribe a minimum of detail, leaving considerable scope for variation in structure and procedures so long as the general objectives of the legislation are being furthered. The means for meeting these goals can be left to WUA members.[15] If some standardization is deemed necessary, strong efforts should be made to retain numerous options such as the number and title of officers, or frequency of meetings, so that WUA members in establishing their constitution or by-laws will be making some choices about how they want to work together. The more decisions they make about such matters, the more they can regard the WUA as "theirs."

Irrigation organizations without legal articles of incorporation exist in many countries and carry out relatively effective water management.[16] The introduction of farmer organizations for irrigation improvement in the Gal Oya scheme in Sri Lanka proceeded without any legal authorization, just administrative support. However, after some years, these WUAs began requesting legal recognition in order to have more standing with those few officials who were still not fully cooperative, to be able to initiate court action against any farmers sabotaging water management efforts, to levy fines if necessary, and to operate bank accounts.

The Philippine experience, according to F. Korten (1982), suggests that devising a legal framework for farmer participation in irrigation management may be a necessary but not sufficient condition for generating a sense of local responsibility for operation and maintenance. Legal aspects of WUAs need to be considered and provided for in program design. The disposition of officials to assume that behavior can be "legislated" needs to be resisted, however. It is preferable to have WUAs which exist and operate first and foremost as social realities, which are reinforced by legal identity and backing. Reversing the sequence -- creating legal entities that then have to gain public support and loyalty -- is less likely to produce the management capacity desired.

Development Strategy

As practitioners begin efforts to establish WUAs, what kind of design strategy is appropriate? There is not a great deal of experience with introducing WUAs successfully, but the record is growing, particularly in South and Southeast Asia. We endorsed a "learning process" approach in the previous chapter when reviewing policy orientations that would best support participatory irrigation management. Here are some of the design implications of this approach:

(1) Although the language of "building" institutions often suggests that WUAs can be "constructed" according to some kind of schedule, program design should remain flexible, and provide enough lead time, so that efforts to develop local institutions are not rushed. In fact, WUAs can sometimes proceed faster than anticipated; progress is not, as the stereotypical view suggests, always slow.[17] However, if WUA management capacities which do not yet exist are called upon, the experience of failure can set back and even abort the process of institutional development. Schedules can be set for purposes of allocating resources, but they must be open to revision, especially to support unanticipated activities and expenditures.

(2) Efforts to develop WUAs should be coordinated with technical modifications in the irrigation system. As suggested earlier, changes in behavior are more likely to be forthcoming when physical changes are also being introduced. Some project designs assign responsibility for farmer organization to certain officials or consultants, often with marginal or uncertain status, and expect them to produce WUAs which will be ready to function at a certain point in time -- the social engineering equivalent of a "turnkey" operation. There is considerable consensus in the literature that such an approach is unlikely to succeed. Without involvement of farmers in design and construction, the physical works are not likely to be as satisfactory (e.g., Illo and Chiong-Javier, 1983), creating grievances among WUA members which can sour their relations with the irrigation bureaucracy. But more important, organizational skills and solidarity are not developed in a vacuum, and engagement of farmers in tasks such as design and construction can be used to lead into greater O&M capabilities (Mayson, 1984).

(3) When introducing a new program, the logic of starting off with pilot efforts should not be undermined by scattering these activities over a very large area, as done in the Sedarhana project

in Indonesia. Although distributing pilot program activities over a large geographic area may be politically desirable, stretching the program's resources too thinly to maintain proper supervision and to assimilate needed "learning" will be counter-productive. In contrast, undertaking activities in a more focused area makes it easier to detect problems and progress and to consolidate learning.

(4) New efforts should probably be targeted first in the relatively more promising areas, where water is neither too scarce nor too abundant for farmers to have incentive to participate in new management responsibilities. One can posit that the incentive will be greatest in the middle reaches of most irrigation systems, and a recent quantified study of farmer participation in irrigation management in Indonesia found this prediction confirmed.[18] Achieving fairly quick and visible results is important for building up momentum behind a farmer participation program, particularly for gaining support from engineers and administrators. It is advisable not to start where farmers are unwilling to assume responsibilities (Lowdermilk, 1985:8-9). It is better to support efforts which can provide examples of effective performance that other users can learn from.

(5) A program should be designed with a view toward stages of development, keeping in mind that some "overlap" is inevitable. D. Korten (1980) has extrapolated from a number of successful rural development programs in Asia, including the NIA program in the Philippines, a sequence for "learning process" programs: first establish what can be "effective," then experiment to see how this can be done more "efficiently," and finally concentrate on devising means for "expansion."

(6) All efforts should contribute toward building up a cadre of personnel who are knowledgable about and committed to a participatory approach. This would include agency staff at various levels, probably with some specialized cadre focusing on farmer organization issues within the agency, as discussed in the next chapter. Under this heading would come also farmer-representatives who are able to articulate the goals and invent the means for better irrigation management. Such persons will be invaluable for mobilizing and maintaining political support for the program, previously discussed as a policy consideration in Chapter 7.

Channels for Implementation

Another key design issue involves engaging agencies and organizations outside of the bureaucracy in the process of developing farmer capacity for irrigation management. Although government agencies are often reluctant to call upon "outsiders" to assist with "their" programs, there are some good reasons why it may be advisable. For example, in this area of activity in particular, an outside agency may help instill the flexibility and normative commitment which bureaucracies generally lack.

Experience with devolving responsibility to private voluntary organizations is not yet very great, but the performance of an agency like CARE in Bolivia shows the kind of contribution which a PVO can make (Lynch, 1983). The fragmentary reports on working with private contractors are not as salutary. They have no incentive to consult with farmers or to accept delays unless their terms of reference are very carefully written and their bases of compensation are set accordingly.[19] Project designers may, with forethought, be able to specify terms and conditions that establish requirements and provide incentives for private contractors to work with farmers so as to increase the management competence of the latter. However, there appears to be little experience with achieving this so far.

One set of organizations which should be considered for involvement in the promotion of participatory irrigation are what can be termed loosely "knowledge-building institutions," especially those with an orientation toward action research. The contributions of the Asian Institute of Management and the Ateneo de Manila University (also later the Ateneo de Naga) to the NIA program in the Philippines were large and important. They contributed valuable and innovative work on "process documentation" and "sociotechnical profiles" (for example, de los Reyes, 1984). The roles of the Agrarian Research and Training Institute and Cornell University in the Gal Oya project in Sri Lanka were also major.

If an agency attempts to introduce more participatory management entirely through the existing irrigation bureaucracy, it is likely to face a hostile or unimaginative reception.[20] Experience suggests that some manner of collaboration with outside institutions can strengthen the role of the most energetic and innovative elements within the irrigation agency. This consideration, however, points to the important subject of how relations between farmers and irrigation bureaucracies can be improved.

NOTES

[1]Farmers within irrigation schemes in the Indian state of Andhra Pradesh are expected to develop by themselves the area below the turnouts being built by the Irrigation Department, constructing field channels and ditches individually or collectively. Officials have reproached farmers for undertaking such land development (the "participation" desired by government staff) either slowly or not at all. Yet a commission of inquiry found that 25 to 40 percent of the designated command area was not or could not be reliably issued with irrigation water due to miscalculations of available supply or faults in design and construction (Hashim Ali et al., 1982). Under such circumstances one could hardly expect farmer organization or participation to thrive. Lowdermilk (1985) gives other examples of technical defects which have impeded farmer participation elsewhere in South Asia.

[2]This issue was discussed at the 1984 Social Science Research Council seminar in Bangalore on "social responses to irrigation" described in the Acknowledgments. There was some consensus that introducing behavioral changes was likely to be more successful in conjunction with physical changes. See report in WAMANA, October, 1984.

[3]Ed Martin and Bob Yoder (now on the staff of the International Irrigation Management Institute) asked leaders of an indigenous WUA in Nepal why it did not undertake other activities like operating a hydro-powered grain mill which would benefit members. The response was, "We have all we can do to manage the irrigation system; there are so many conflicts." A grain mill has now been installed in that community, however, and appears to be operating satisfactorily (personal communication).

[4]Cernea says that the WUAs in South Korea have apparently benefited from a history of nearly two generations of prior cooperation. It should be noted also that these groups are quite regimented and regulated, so similar results should not be expected unless a government is willing and able to operate as intrusively as reported by Steinberg et al. (1980).

[5]This does not constitute an endorsement of the T&V system, which has attracted criticism as well as support in the countries where it has been introduced (Moore, 1984). Working with and through WUAs which are accountable to farmer-members would, however, mitigate one of the common problems with the T&V approach, that "contact farmers" are not necessarily responsive to the needs and ideas of their fellow farmers.

[6]This system was able to almost double cropping intensity in just two years' time (Kathpalia, 1984).

[7]Plusquellec and Wickham (1985) suggest from their understanding of Thai experience that 125-150 acres represents the maximum size of primary service area for good irrigation management. "There is evidence that many service areas are too large to permit timely irrigation throughout their commands, with the result that water is wasted and yields reduced" (1985:49). Recall that in the Nong Wai scheme in Thailand noted above, before a system of WUAs was introduced, primary service areas were 10 to 12 times larger than this maximum. Plusquellec and Wickham suggest that 20-25 farmer-members represent an optimum size of membership, although we favor somewhat smaller base groups. The Nong Wai rotation groups were one-half to one-quarter as large as the "optimum" Plusquellec and Wickham recommend.

[8]Maass and Anderson (1978) consider that the community of irrigators is best regarded as a group of persons who have shared rights to the use of a common water source, emphasizing legal or customary rights rather than ownership or use of land that is irrigated. In the case studies, we found such rights, with few exceptions, to be attached to land or to membership in some community, residential or kinship. An exception would be the zanjera groups in the Philippines, where investment of labor and funds earned members a share (atar) of water independently of land or residence.

This case raises a qualification to our conclusion, as the zanjera organizational form is made appropriate by the fragmentation of landholdings within these systems. To have had WUAs of field neighbors would require most farmers to be members of several WUAs and to have many separate work obligations. Where there is significant fragmentation, the size and structure of WUAs must be adjusted to suit this circumstance.

[9]"Experience shows that while women participate actively in the [WUA] meetings, generally they are not formal members because it is assumed that each household will have only one member, in which case it is normally the man. This means that women cannot be officers of the association -- even though often they are well qualified. Mechanisms to allow joint memberships from each household have been discussed to avoid this problem." (F. Korten, 1982:16). In San Pedro de Atacama, Chile, a similar situation prevailed where women participated in meetings, but did not assume positions of responsibility largely because men were the "members" (Lynch, 1978).

[10]In Bangladesh where landlessness is a widespread and serious problem, one experimental solution has been to establish not WUAs but WPAs (water producer associations) by loaning money to groups of landless so they can purchase pumps and operate shallow tubewells, selling water to landowners and cultivators (Wood, 1984). This is an innovative approach, and although not always successful it warrants further experimentation.

[11]A contrary suggestion, that WUA decision-making be entrusted to persons acting in executive rather than committee or assembly roles, was made by Chambers (1975) based on limited interviewing of farmers in Sri Lanka. Other data gathering and analysis in Sri Lanka has provided evidence of farmers favoring more participatory roles in grassroots irrigation management (Uphoff et al., 1981).

[12]A methodology for mobilizing constructive new leadership is outlined in Uphoff (1986). Conscientious leaders are like rice plants, productive but tender and vulnerable, whereas self-aggrandizing leaders are like weeds, which will crowd out the "rice plants." If there are enough good leaders, they can hold the undesirable ones in check, but the latter can take over a whole program once they reach a certain threshold, not necessarily a very large number. Organizers who identify and encourage constructive leaders are like rice farmers who prepare the seedbed and help productive "plants" get established, keeping "weeds" in check so they do not take over. This said, it is crucial that the leaders be selected by and fully acceptable to the broad majority of farmers and not be those farmers who are most attractive to outsiders.

[13]The Boards were required to: (1) manage the delivery of water; (2) develop a plan for operation, maintenance, improvement and rehabilitation of the watercourse; (3) supervise construction and maintenance of the watercourse and other improvement activities; (4) employ and discharge ditchtenders, collectors and construction personnel; (5) exercise emergency powers to repair watercourse breakages; (6) negotiate and contract with government agencies and other institutions for improvement programs acceptable to the general assembly; (7) serve as the communication link between government agencies (and WUA members) in dissemination of information and all matters representing the views and requests of the irrigators; (8) maintain the financial and organizational records of the association; and (9) call special meetings of the general assembly for any matter involving original expenditure and other important issues involving the general membership (cited in Cernea, 1984).

[14]The Pakistan act discussed above provided fourteen "powers of association" which are almost all obligations: (1) improve, rehabilitate, operate and maintain the watercourse, (2) improve the water supply, (3) locate, own, operate and maintain tubewells and lift pumps, (4) sanction upgrading and maintenance of farm ditches and field outlets, (5) encourage adoption of improved on-farm water use and management practices, (6) participate in programs to improve watercourses, land leveling and agronomic practices, etc., (7) establish water delivery schedules and supervise water allocation and distribution "in such a manner so as not to interfere with canal water delivery," (8) set and collect general and special assessments, (9) conscript labor for emergency repairs on watercourses, (10) locate, install and maintain drainage facilities, (11) remove obstructions in water-courses, (12) enter into contracts for obtaining loans and grants and setting a repayment schedule, (13) ensure that all members' rights under the law are respected, and that each member gets his fair share of water in a timely fashion, and (14) ensure that all members contribute their fair share of labor, money, etc. Sanctions may be imposed if these terms are not carried out. Note that there is no mention in the act of any obligations placed upon the irrigation bureaucracy to meet any performance standards vis-a-vis water users. Essentially the same irrigation ordinance was proposed for Sri Lanka, enumerating "powers" that were really duties. Fortunately its enactment was deferred pending experience gained with a more "bottom-up" approach (Uphoff, 1982).

[15]In Indonesia, where there is a high turnover of cultivators -- as much as 50 percent per season, due to high rates of tenancy -- it is difficult to establish uniform membership and duties. The government set up a legal standard of membership in WUAs, but farmers prefer to think in terms of tasks, guided by a principle of equity. As noted above, farmers are expected to participate in certain kinds of work or decision-making depending on their land tenure status and the nature of the task. Work that contributes to long-term improvement is done by owners only; all cultivators are obliged to help in activities that improve irrigation for a particular season. WUAs also consider how rich or poor a person is. Someone who they consider cannot afford the outlay may be excused from having to contribute (John Duewel, personal communication). WUAs may make some such adjustments to suit local conditions and norms no matter what the law says, but it is better if they need not get unnecessarily embroiled in legal controversies.

[16]Cernea (1984) cites a study by D. Craig Anderson, "Irrigation, Institutions and Water Users in Ecuador" which found

no difference in the water management performance of WUAs that were legally constituted and those that were not.

[17]In the Gal Oya case in Sri Lanka, initiatives respectively by the farmer-representatives and the top civil servant in the district established third and fourth tiers of organization before the second tier had been set up by the program.

[18]Robinson (1985) measured farmer participation in design, construction and maintenance for a dozen small-scale systems being (or having been) rehabilitated. Over 700 farmers were interviewed in this study. The level of participation in these three activities plus formal participation in WUA activities was higher for farmers in the "middle" of these systems than for farmers in the "head" or the "tail." Higher participation was statistically significant at the 5% level for design and maintenance. In Gal Oya, although organizing activity had to begin at the "head" in order to be synchronized with rehabilitation planning and implementation, experience showed more rapid "take-off" of farmer participation in the "middle" reaches.

[19]Experience from Indonesia and Sri Lanka was reported at a May 1984 workshop at Cornell with difficulties in getting contractors to pay attention to farmers and to their ideas and needs.

[20]There are some examples where participatory management has been introduced from within the irrigation bureaucracy. A good example is the Minipe scheme in Sri Lanka, where a Deputy Director of Irrigation (now Chairman of the Mahaweli Engineering and Construction Authority) took such an initiative. In this case, however, in the initial stages, he received the assistance of a PVO, the National Heritage Association, which provided young volunteers to work as organizers. In the Nong Wai irrigation scheme in Thailand, the program of introducing WUAs was handled by a combination of government agencies, though with outside consultants from the Asian Development Bank (Kathpalia, 1984).

9
Improving Agency Relations with Farmers

Mutual complaints between farmers and agency personnel are endemic in irrigation management. Establishing the validity of grievances is difficult because of their complexity and long-standing nature. Determining facts is elusive because of the overlay of negative attitudes commonly found on both sides. Program designers and managers intending to increase farmer participation should be mindful of the history of relations between farmers and officials, but little will be gained by trying to render summary judgments on past practices. The best approach is to encourage both sides to seek a new start in their relations, recognizing that both can gain from better irrigation management which includes farmers' involvement. To the extent that farmers and engineers are willing to look self-critically at their own past performance, of course, the prospects for future cooperation are improved.

There are two "models" for participatory irrigation management. One is a "division of labor" approach which negotiates and maintains separate, though still interdependent assignments of activity; the other is driven by a concept of "collaboration." The first tends to operate in "zero-sum" terms, where the gain of one is the loss of the other, while the second is more "positive-sum." Which is more appropriate is influenced by the quality of relations prevailing and by the nature and level of the task at hand. When negative opinions are strong, the first may minimize conflict and be the best arrangement. But relations can change, and as more mutually respectful attitudes arise, collaboration may emerge. Certain activities like maintenance and drainage are more amenable to a division-of-labor approach, which is also likely to be more relevant at lower levels of a system than at higher ones. Hence some combination of both approaches will generally be appropriate.

In any irrigation system where agency personnel are involved in management, it is important that the main system managers accept the legitimacy of farmer participation if this is to become effective. Since organizing for collective action is itself a demanding and problematic undertaking, water users can be fairly easily discouraged from investing in developing group capacities, thereby reverting to or remaining in an anarchic situation. Further, there needs to be reliable main system management for farmers to be able to make and carry out plans for activity at their respective levels (Wade and Chambers, 1980). If main system management is itself unpredictable, this encourages what Hart (1978) has graphically described as "anarchy under the canals."

The theme of "bureaucratic reorientation" was introduced in the preceding chapters. Any government agency, not just one involved in irrigation, will have certain professional self-images, standard operating procedures, institutional doctrines, typical incentives and career patterns that shape the behavior of its personnel. To the extent that these lead technical staff to adopt condescending attitudes, to deprecate farmer's ideas and overlook farmer's needs, to resort to coercive means for gaining compliance rather than seek understanding and agreement, the aim of engendering more participatory irrigation management will be thwarted.

Paradoxically, the stronger and more self-confident an agency is, the less receptive it may be to new modes of operation. On the other hand, a weak agency, while receptive, may have difficulty in winning cooperation and budget support from other arms of government. The challenge of bureaucratic reorientation is considerable; however several recent experiences show promise in this regard. The following discussion reviews some of the elements that can contribute to a "new start."

Agency Incentives

Organizations do not modify their objectives and practices very readily. We have found three major kinds of incentives which can encourage an agency to rethink and redirect its activities: (a) failure of conventional approaches, (b) resource imperatives, and (c) inter-agency competition. It is hard for agencies to recognize, let alone admit that they are not fulfilling their mission adequately. But such a realization can open bureaucratic ears and eyes to new approaches. Visible deterioration or errors in placement of structures can be an embarrassment and give impetus to work in new ways which correct or avoid previous problems as these become evident. When farmers in numerous Philippine systems simply filled in and cultivated over many of the field channels the National Irrigation Administration had built without consulting them, it was evident that NIA was not meeting farmers' needs and was wasting resources (Illo and Chiong-Javier, 1983). Unfortunately, similar experience in the government-initiated Kosi irrigation project in India did not induce the irrigation agency there to embark on the same kind of internal reform as NIA undertook. Pant describes the following situation, which appears ripe for bureaucratic reorientation.

> All the field channels constructed were earthen work of very sub-standard nature, and were cut or obliterated invariably by farmers. The coverage of 44,000 hectares of land with field channels is an impressive figure, but the way it has been done and the hurry in which it was done indicates that a ritual has been observed. (Pant, 1981:A81)

As noted above, a strong and self-confident agency may require a string of setbacks before it is willing to think it might do better by working more closely with water users. Technical shortcomings may only become apparent when faced with other sobering problems, such as a need to obtain budgetary resources or to counteract bureaucratic challengers.

The National Irrigation Administration in the Philippines embarked on its innovative course after the government directed that agency to begin recovering the costs of its capital investment in improving communal (small-scale, user-managed) irrigation schemes. The leadership of NIA realized that farmers would not accept repayment unless they had some voice in whether (and what kind of) structures would be built on their behalf. In contrast, the design for the Gal Oya project in Sri Lanka originally assumed (without any prior consultation) that farmers would do all tertiary rehabilitation on an unpaid basis (this was someone's idea of "farmer participation"). The fact that no funds had been budgeted for this work gave engineers an incentive to begin meeting with farmers, to discuss plans for rehabilitation and to adjust these to field-level problems, since farmers could easily withhold voluntary labor if they were not satisfied with the plans. Rehabilitation at primary and secondary levels would produce little benefit unless there was tertiary renovation as well. What appeared at first to threaten the participation of farmers (the imposition of "voluntary labor") unexpectedly created conditions for encouraging their participation, once engineers approached them in a respectful and cooperative manner.

Bureaucratic competition can also create incentives to innovate. If an agency sees some other agency forming linkages with "it's" client group, this may spur the agency to form its own links to preempt the loss of contact and support. The program for establishing Command Area Development Authorities (CADA) in India has created some impetus for line bureaucracies to rethink their strategy. In the Philippines, after the electrification agency started using organizers to establish irrigator groups around a common pumpset, NIA contracted with the Farm Systems Development Corporation to start Irrigator Service Associations linked to NIA. The rapid rise and bureaucratic threat of the Mahaweli Authority in Sri Lanka threatened the Irrigation Department there, which had primarily concentrated on design and construction. If O&M was to become a more important part of the Department's operations, it would need better cooperation with farmers.[1]

Presumably one does not want to have or to encourage "failure" as an inducement for bureaucratic reorientation. On the other hand, agencies often ignore and deny evidence of shortcomings in their conventional mode of operation. By encouraging candid self-assessments, a climate of opinion may be created wherein leadership for change comes from within the agency.[2] Creating a certain amount of "dependence" of the agency on the resources of water users can have a salutary effect on its attitude toward farmers, though it is preferable in the long-term to seek interdependence marked by cognitive respect on all sides.[3] Deliberately fostering bureaucratic competition is a risky strategy, but it has often been the surest spur to improving performance.

Technical Cooperation

One "division of labor" approach to farmer participation is to assign to water users certain responsibility for organizational matters, reserving all technical decisions and evaluations to the irrigation agency staff. Some staff may try to protect their "turf" by exaggerating the complexity of technical tasks (and maybe the extent of their own expertise) or by deprecating the knowledge of farmers. This is unfortunate because water users usually have a great deal they can contribute to design, construction, operation and maintenance of irrigation systems, not just in terms of labor but also ideas and innovations.

Various examples of "dams that failed" were cited in Chapter 7. Documenting and publicizing such outcomes may make engineers a little more willing to listen to farmers "next time." The conclusion should not be that technical personnel are incompetent but that engineering and agronomic judgments are complex and imperfect. No one should expect technicians, however highly qualified, to be correct all the time, given the levels of precision needed for the most efficient irrigation.[4] By involving farmers in gathering and assessing information, and in evaluating technical alternatives, the chances of making unfortunate decisions should be diminished. If decisions subsequently turn out to be wrong, farmers cannot simply "blame the engineers," which is easy to do if they have had no role in thinking through the complexities and alternatives.

To the extent that the structures designed and built are inappropriate, the prospects for getting farmers to take responsibility for their operation and maintenance are diminished. Trying to get users to manage, let alone pay for such a system becomes a source of conflict.[5] On the other hand, if farmers are involved in planning and are able to make suggestions, this conveys a sense of respect for them and a feeling of common purpose that can buffer the disappointments over shortcomings in design and performance that are nearly inevitable in irrigation.

Engineers need not and should not surrender their best technical judgment when dealing with any problem requiring scientific expertise. They are expected to contribute the benefits of their learning and experience. But these can usually be applied to better effect in conjunction with the fruits of users' observations and experimentation (Chambers, 1983:82-101). It will be a great step forward toward promoting farmer participation if agency personnel will recognize that technical problem-solving can be a joint enterprise.

Engineers and farmers should be able to contribute from their respective funds of knowledge to the formulation of proposed solutions, remembering that they require experimentation and evaluation. Indeed, one indicator of bureaucratic reorientation is the degree to which agency personnel do not see working cooperatively with water users on technical problems as depriving themselves of status. In a reoriented agency, this would become a source of satisfaction.

Special Roles

Where relations between engineers and water users are estranged, as is often the case, improving them will probably require introduction of some number of persons working in specialized roles to build up or assist water user associations. Such organizers, sometimes referred to as "catalysts," would be recruited on the basis of personal qualities that make them good intermediaries -- among farmers to encourage them to participate in WUAs, and between farmers and officials to establish "vertical" cooperation. The NIA experience in the Philippines is the best documented example of this (F. Korten, 1982; Bagadion and F. Korten, 1985), but the same methodology has contributed to improved relations and performance on the part of both farmers and engineers in the Gal Oya scheme in Sri Lanka (Widanapathirana, 1984; Wijayaratna, 1984; Uphoff, 1985 and 1986).

There are instances where positive changes have been introduced by technical personnel, e.g., the Minipe and Kimbulwana schemes in Sri Lanka (de Silva, 1981 and 1984; Weeramunda, 1985) and the Nong Wai system in Thailand (Kathpalia, 1984). In Minipe, however, there was assistance in the early years from volunteer organizers provided by a Buddhist voluntary service organization and Kimbulwana was a small scheme, inspired by the first. In the Nong Wai system, water user groups had previously been formed at the direction of Cooperative Department officials, but little had been done "to activate them." In 1981, Asian Development Bank consultants were brought in to help bring about more productive relationships among farmers and with the Royal Irrigation Department (RID). The RID had a special O&M unit within the agency which had responsibility for farmer organization, and staff from this unit were assigned to work with the consultants. Together they began to work with a larger group of RID employees who had responsibilities at the field-level.

> The staff was trained to be persuasive and instructive and not to have an authoritarian attitude. Once farmers' confidence was gained through open and frank discussion by explaining the reasons for every suggestion and changes made, their cooperation was forthcoming. The process was slow but succeeded, more so than a directive approach (Kathpalia, 1984:18).

An interesting aspect of how RID staff were able to change their traditional modus operandi and work more cooperatively with farmers involved reducing the size of the territory for which they were responsible by one-third, to increase the intensiveness of their work. Rather than try to take over the responsibilities of the Cooperative Department, the RID worked in a complementary and supportive role to increase its capabilities for dealing with the organizational issues of participatory water management. This effort was expanded to the whole project area of 25,000 acres, with 5,500 farmers organized into 169 groups handling O&M for 248 basic operational areas (chaks) averaging about 100 acres each.

A cadre of organizational specialists for promoting participatory management may not be needed if the irrigation scheme is fairly small, such as Kimbulwana (1300 acres), or if the technical leadership possesses unusual qualities of personality and value commitment as was the case in Minipe.[6] The responsibility for establishing and sustaining farmer organization can be delegated to regular irrigation department staff, provided they are given appropriate training, incentives and opportunity, especially enough time and resources to work on these tasks. It must be remembered that technical staff are recruited for their engineering or other technical qualifications, not for their interpersonal skills or understanding of organizational dynamics. They are usually overburdened with existing tasks, so that added responsibilities may get short shift, especially if these are duties they feel somewhat uncomfortable with.

During the process of project design, it may be possible to reorient technical personnel to work more productively with farmers. Examples of this include the work undertaken in Nong Wai and that attempted (not very systematically or successfully) in the Pochampad irrigation system in India (Singh, 1984). We did not find cases where agricultural extension personnel or an "extension" approach were used with good results, but we cannot say this is impossible. Extension operations tend to be a matter of "telling farmers what to do." This is unlikely to be effective for producing better local capacity for water management. If the relationship with farmers is more one of consultation and collaboration, there is a greater possibility for efforts to succeed.

The most widespread and satisfactory results thus far have been with specialized roles placed within or alongside the irrigation bureaucracy. Such catalysts work with farmers and with engineers to establish new attitudes and new ways of interacting. Data from the Philippines and Sri Lanka on respective costs and benefits, reported in Chapter 2, suggest that the costs of using organizers can be recouped as quickly as within two years, after which time a valuable water management capacity should be sustainable with only modest continuing expenditures.

Unfortunately, social infrastructure is like physical infrastructure in that it requires some maintenance investments. Even after WUAs have been established, there is need for training of new farmer-representatives, for helping to resolve problems that arise within and between WUAs and that they cannot handle themselves, and for sorting out farmers' difficulties with the bureaucracy in an "ombudsman" role. If engineering staff cannot handle such duties, or if this is not a good use of their time, there is reason to have a special cadre of persons who can deal effectively with "institutional hydraulics." Clear responsibility and adequate means should be given so that this support function is not neglected. Any specialized cadre should be fully part of the agency, so that socio-technical syntheses are continually made.[7]

Agency Style

The way in which an agency goes about its business has effects apart from what it does. One of the truisms of organization theory is that organizations tend to replicate externally the kinds of relationships and values they display internally. If an irrigation department is rigid and hierarchical in its own dealings, it is likely to encourage similar patterns in the groups with which it works. If it wants WUAs that are efficient, participatory, and responsive, it needs to set such examples for them in its own performance.

It would thus be inconsistent, and likely ineffective, for agency leadership to "order" its staff to work cooperatively with water users. Movement toward a more participatory approach to irrigation management should be the result of discussions and experimentation. It should involve farmers in ways that encourage new attitudes and behavior on the part of engineers and technicians, and corresponding changes among farmers themselves in an iterative process.[8]

Certainly "leadership" within the agency is important. Both the Administrator and especially the Deputy Administrator within NIA in the Philippines gave support and direction to the efforts for farmer participation.[9] In Sri Lanka, initial support came more from the parent ministry than from the Irrigation Department itself, though several of its Deputy Directors have given leadership within their sphere of responsibility. Unfortunately, little is written on this subject in the literature, so we cannot comment on it more systematically even though it is clearly an important factor.

Agency Organization

In addition to agency style, the structure of an agency is important. For example, real barriers for participation arise from having a sharp division between the processes of administration, staffing, budgeting, communication, etc. for design and construction, on the one hand, and for operation and maintenance (O&M), on the other. Unfortunate design decisions are often made without regard to their implications for O&M. Since design and construction yield more professional rewards in terms of prestige, promotion and earnings for engineering staff, O&M, which is of most immediate concern to farmers, gets treated like a stepchild. Supposedly the best staff are assigned to design, so engineers given O&M duties often seek transfers out (unless there are economic, often illegal, compensations adhering to O&M). Staff turnover obviously disrupts contact with WUAs (and corrupt practices corrode such contact). Reducing the dichotomy in staffing, prestige, etc. between design and construction and O&M (a worldwide phenomenon) can contribute toward better working relations with farmers.

One organizational initiative could be to establish a high-level working group to oversee the process of introducing participatory irrigation management as was done in the Philippines when NIA set up the Communal Irrigation Committee discussed in Chapter 7. This brought together top professionals in engineering and administration with consultants from the donor agency (the Ford Foundation) and social scientists assisting in the process from their base in knowledge-building institutions such as the Asian Institute of Management, as discussed in Chapter 8. This group received regular reports (process documentation) on what was going on in the field and on what was being learned in the "learning laboratories" that the project maintained (D. Korten, 1982; F. Korten, 1982). It was in a position to give informed advice and to sustain the support needed for the effort from many sources.

The possibilities and appropriate strategy will be different for each country. In Sri Lanka there was no working group like the Communal Irrigation Committee in the Philippines because there was no initial top-level support within the Irrigation Department for any significant farmer role in irrigation management; farmers were expected simply to obey the ID's directions. It did however accept some experimentation which the Ministry of Lands and the donor agency (USAID) supported. An informal network of key individuals in several ministries, departments and institutes played a role equivalent to the Communal Irrigation Committee's.

Fairly rapid reorientation is possible with tailored and persistent efforts as can be seen from the amount of change in Sri Lanka. Within two-and-a-half years, the top echelons of the Irrigation Department approved a proposal for irrigation management reform which included a four-tiered system of farmer representation with participation in management right up to the main system level. (There was some bureaucratic pressure on the ID to come up with such a proposal, but the recommendation was reportedly "unanimous.") Farmer organizations, although only tolerated at the outset, became accepted as management partners by the technical staff in the field.[10]

Concluding Observations

In the next ten years, numerous other countries will have some need for transformation in the orientation of their national irrigation bureaucracies. This will take many different forms and will produce a plurality of results. The case for more participatory irrigation management appears stronger each year, as new positive results are registered and as the deficiencies of conventional approaches become more evident. The recent dates of much of the literature in the Bibliography indicate how rapidly the field of knowledge is expanding.

The growing "fiscal crisis" in LDCs is putting ever more budgetary pressure on governments to find less costly ways of providing services. Yet governments cannot expect or require farmers to contribute more resources in the form of labor or repayment of construction costs without giving them some greater role in decision-making. As seen in Chapter 3, although irrigation management activities are distinguishable analytically, they are linked multi-dimensionally, and it is unrealistic to isolate just one or two functions for farmers to perform or to be involved with.

This review of experience comes at a time when experimentation with participatory irrigation management is expanding. Attitudes which were fairly fixed until recently are now open to new evidence. An international expert consultation on water management organized by FAO and USAID and held in Indonesia in July 1984 produced remarkable consensus on these new directions.[11] Indeed, one of the chief problems confronting professionals seeking practical advances of a participatory nature is no longer resistance from main-line agencies but rather some of the claims and expectations emanating from over-enthusiastic proponents of new approaches.

This is an area where the welfare of hundreds of thousands, even millions of persons is at stake. Improvements in the efficiency and reliability of irrigation however achieved will contribute greatly to household, regional, national and international objectives. Increased appreciation and support for an expanded user role in irrigation management appears to offer one of the most beneficial and least costly avenues for such improvements. This is important as governments contemplate the rising cost of expanding new irrigated acreage and their shrinking resources for such investment.

The participatory approach is not without its disappointments and difficulties. There are no general "blueprints" which are useful for all locations and all circumstances. The orientation of the irrigation agency is the key variable. Evidence continues to accumulate of farmers' willingness and ability to discharge greater responsibilities. After we had selected our initial 50 cases for analysis, we kept coming across new reports of encouraging experience where users had been involved more actively in water management, for example, in Thailand (Mayson, 1984), Bolivia

(Lynch, 1983) and Senegal (Cernea, 1984). In the latter case, a World Bank document says that once the irrigation agency (SAED) formed water user groups of 12-15 family heads each, these groups did "a better job than SAED" in maintaining agricultural equipment to improve production.

The principal question is whether engineers and officials are willing and able to depart from technocratic and paternalistic postures, to begin working more as partners with those for whom the irrigation enterprise is undertaken -- the water users.

NOTES

[1]In the U.S., when the Soil Conservation Service set up farmer groups, this was an impetus for the Army Corps of Engineers and the Bureau of Reclamation to improve their respective links with farmers, if only to gain and maintain political support, as was pointed out at our May 1984 workshop.

[2]In the NIA case, the U.S. aid agency provided foreign travel and training opportunities for a number of rising young engineers in the late 1950s and early 1960s. Their opportunity to observe irrigation systems particularly in Taiwan and U.S. where farmers had major management roles got them thinking about how to improve irrigation performance in the Philippines, according to former NIA Deputy Administrator Benjamin Bagadion (seminar presentation, Cornell University, June 1, 1985).

[3]A good example of project design creating "dependence" on farmers was the rural roads component of the PIDER rural development project in Mexico. A special office for labor-intensive road construction was set up within the Ministry of Works. Because it was given little heavy machinery and no funds to pay for outside labor, it had to rely on rural communities to achieve its bureaucratic goals. Road committees were formed in the villages eligible for assistance, and plans for new or improved roads had to be worked out jointly with the communities. Engineers knew that if their plans were unacceptable, no local labor would be forthcoming. This component turned out to be the most successful part of the project, building or upgrading 65,000 kilometers of rural roads in a five-year period (Edmonds, 1980). The participatory methodology developed in the PIDER project is reported in Cernea (1983).

[4]Our colleague Gil Levine has pointed out that differences in elevation of as little as 5 cm. can have a large impact on the performance of gravity-flow systems for rice cultivation. Most topographical surveys in LDCs have margins of error many times this much, and the costs of getting greater accuracy are utterly prohibitive. It should not be surprising that in one large irrigation system in a Southeast Asian country, fully 40 percent of the field channel outlets were sited incorrectly by solely "technical" procedures, without field testing and without consultation with water users (Wade, 1981).

[5]This discussion has focused only on inadvertent errors, leaving entirely aside the possibility and enduring problem of dishonest performance. According to Rao (1984): "The leakages [in executing project works] are generally estimated to be around one-third of the official budgets.... As a whole, it appears that only about half of the magnitude of the officially estimated costs should be taken as real costs" which would provide a fair basis for calculating charges to farmers for capital repayment or operating fees.

[6]The engineer responsible for introducing the new approach to water management at Minipe has reported that he got better, i.e., longer-lasting, results in the pilot area where he was assisted by catalysts, however (de Silva, 1984).

[7]One of the most successful cases of developing local capacity, not in the irrigation sector but in an equally technical one, is the Kenya Tea Development Authority. It works with smallholders, thought by the colonial authorities to be incapable of growing high-quality tea but who have now been able to surpass the estate sector in quality and efficiency of their tea production. This has been made possible in part by the establishment of a network of grower committees, vertically linked in a four-tiered organization much like in an irrigation scheme. The KTDA created at the outset a special division of its organization to set up and support the system of farmer committees (Lamb and Mueller, 1982).

[8]When the ARTI-Cornell team first visited Gal Oya in January 1980 to begin planning an experimental program for farmer participation, it found relations between farmers and engineers quite estranged. Most of the farmer malpractices which had contributed to the deterioration of structures and operations at lower levels were traceable to poor management of the main system. The team concluded that unless and until engineers changed their attitudes and behavior, one should not expect

farmers to change theirs. As it turned out, with some small but constructive and mutually supportive changes on both sides, a process of improving relations -- and system performance -- was initiated. Indeed, the initial conclusion was wrong because, as with so many things, the change could not be one-way. Even small changes from the ID side encouraged farmers to take more responsibility, which challenged engineers' negative stereotypes about farmers and evoked more positive orientations from them.

[9]The Deputy Administrator Ben Bagadion has been referred to several times above. The Administrator wrote in an editorial for the Philippine Agricultural Engineering Journal, 10(2), 1979: "Local organizations are the crying need of rural development if our rural people are to play a more vital role in the country's development...Assistance to communal irrigation...certainly deserves the government's strong commitment and support."

[10]This was stated by the current Director of Irrigation for Sri Lanka at a national workshop held in May 1986 at the International Irrigation Management Institute, Digana Village, Kandy, Sri Lanka.

[11]The final report is available from the International Support Programme for Farm Water Management, Land and Water Development Division, FAO, Rome. The case studies are published in FAO (1985). For an earlier summary of that Progamme's conclusions about participatory irrigation management, see FAO (1982), which includes brief case studies from Sri Lanka, Indonesia, Pakistan and the Philippines.

Annexes

Annex 1

Criteria of Irrigation Management Performance

Many variables and criteria have been suggested for analyzing and evaluating irrigation water management. Because there is such multiplicity, we have tried to organize them into a coherent set of variables, integrating criteria commonly referred to in the literature. One can start with the analysis by Mary Tiffin (1983) done for the Irrigation Management Network of the Overseas Development Institute of London. She identified thirteen "attributes of water supply systems." Some of these, however, were opposite values of the same variable, e.g. Fixed Availability, and Flexible Availability; or Management Adjustable, and Farmer Adjustable.

Most of Tiffen's criteria can be encompassed within three general variables. Supply (I.) is the principal variable, with the criterion of Adequacy (I.A.) summarizing many other characteristics as discussed on page 183. Flow (II.) and Distribution (III.) constitute the other two major dimensions for judging how well irrigation systems perform.

I. **SUPPLY**

A. ADEQUACY -- amount is sufficient to meet crop requirements

B. TIMELINESS -- amount is delivered when crop requirements occur

C. RELIABILITY -- amount is delivered when expected and needed; contributes to predictability (II.C.)

II. **FLOW**

A. VARIABILITY -- ranges from:

1. Steady flow -- constant or near constant, to

2. Fluctuating flow -- which may or may not be adequate, timely, reliable, predictable, etc.

Steady flow (II.A.1.) may meet the need for timeliness (I.B.) and reliability (I.C.) but not always adequacy (I.A.).

B. FLEXIBILITY -- ranges from:

1. Adjustable timing -- no limitations on period in which water can be made available, to

2. Fixed timing -- strict limitations on period of delivery; supply (I.) can vary within this period.

C. PREDICTABILITY -- ranges from:

1. Regular -- water flow, whether steady or fluctuating, is known in advance, to

2. Erratic -- flow is either (a) not known, or (b) not knowable.

Regularity (II.C.1.) permits users to take steps to correct deficiencies if supply not adequate or timely. It also permits users to make better use of complementary inputs.

III. **DISTRIBUTION** -- varying in terms of:

A. CONTROL

1. User management,

2. Joint management, or

3. Agency management.

B. EQUITY -- extent to which access to water is equal, according to some criterion of:

1. Area to be served,

2. Crop to be served, or

3. Persons to be served.

This is likely to vary by location (between head and tail), but can be affected by land tenure status or other variables.

C. RIGHTS -- claims to water can be based on criteria of:

1. Time -- first in time; or prior appropriation,

2. Location -- upstream access; or riparian rights,

3. Other criteria, e.g., prior investment in developing water source, beneficial use, or status user's in community.

ADEQUACY (I.A.) is affected by all the other variables:

TIMELINESS (I.B.) -- supply cannot be adequate (I.A.) even if the amount is ample when the timing is wrong.

RELIABILITY (I.C.) -- unreliable supply is also unlikely to be adequate (I.A.).

VARIABILITY (II.A.) -- steady flow (II.A.1) is preferred, though not if this is inadequate; a fluctuating flow (II.A.2.) may or may not be adequate.

FLEXIBILITY (II.B.) -- adjustable flow (II.B.1.)is preferred, presumably up to point of adequacy; fixed timing of flow (II.B.2.) may or may not be adequate.

PREDICTABILITY (II.C.) -- if one can predict an inadequate supply, efforts can be made to compensate by getting more water (supply) or by changing the cropping pattern (demand).

CONTROL (III.A.) -- users (III.A.1.) want as much control as possible to deal with any inadequacies of timing, reliability, variability, flexibility, predictability, etc.; agency managers (III.A.3.) will be similarly motivated if they believe they can compensate for these inadequacies better than can users; some arrangement for joint management (III.A.2.) may best deal with problems of supply.

EQUITY (III.B.) -- this is ultimately a test of adequacy (I.A.), since an inequitable distribution will be inadequate for at least some users unless supply is always abundant.

RIGHTS (III.C.) -- when not every user can attain adequacy (I.A.), there must be some system for allocating water; systems of rights establish claims for distributing water when its supply is in any way inadequate, untimely or unreliable.

Clearly water supply is a multi-dimensional phenomenon. Users will often be in a better position than agency managers to make adjustments to optimize amount, timing, flow, and distribution. However, they need to have some foreknowledge of supply characteristics so as to be able to make adaptations in delivery to meet contending objectives.

Annex 2

Analysis of Irrigation Activities

Hunt and Hunt (1974) offered the most detailed initial analysis of irrigation management activities to be found in the literature, but they have some omissions and underemphases because they did not map out activities along dimensions as we did. They discuss allocation, construction, and maintenance, giving attention also to decision-making, resource mobilization and management, and conflict resolution. However, distribution, operation and communication (the third activity in each of our three sets) are treated only implicitly. Acquisition is discussed in terms of "rituals ensuring water supply" and also "defensive warfare" to protect supply. They do not deal with acquisition or design as specific tasks, and drainage is ignored.

Coward (1979) in his treatment of the subject suggested water allocation, system maintenance, and conflict management as the "fundamental tasks" of irrigation. These three activities deal respectively with water, structures, and organization. He subsequently added acquisition of water and resource mobilization to the list. Coward participated in our working group discussions and concurs in this expanded framework for analysis of activities. We note also the listing of irrigation management activities by Kelly (1983), and Freeman and Lowdermilk (1985).

Activities	Hunt and Hunt	Coward	Kelly	Freeman and Lowdermilk
Acquisition	Rituals, defensive warfare	Acquisition of water	Control of water source	—
Allocation	Allocation	Allocation of water	Allocation of water	Allocation
Distribution	—	--	Delivery of water	—
Drainage	—	—	Drainage	Drainage
Design	—	—	—	—
Construction	Construction	—	Construction of facilities	Construction
Operation	--	--	Operation and maintenance	—
Maintenance	Maintenance	Maintenance of system		Maintenance
Decision-Making	Decision-making	—	—	--
Resource mobilization	Resource mobilization and management	Resource mobilization	—	—
Communication	--	—	—	—
Conflict management	Conflict resolution	Conflict resolution	Resolution of conflicts	Conflict resolution

In our first formulation of the framework, we listed evaluation as one of the generic organizational activities applying to acquisition, allocation, distribution, etc., and did not include communication. When we shared a first draft of our framework with Robert Chambers, one of the most thoughtful contributors on the subject of irrigation management, he suggested by return mail that we subsume evaluation under decision-making, where it could be fitted conceptually, and that we add communication -- "so universal that one doesn't see it" (personal communication, 23 April 1984). In fact, by the time we received his letter, we had ourselves concluded the same thing and had already made this change. Note that none of the three analytical schemes on the previous page list "communication" as an essential activity in irrigation management, in line with Chambers' observation.

Readers who are familiar with the structural-functional analysis proposed by Talcott Parsons will see some similarity between his formulation and ours, though our working group's discussions of irrigation experience, did not presume any prior theoretical categories. We started without any intention of paralleling or of re-inventing Parsons' scheme, which proposed the following four "functions" as common to all "systems."

Functions (Parsons)	Activities
GOAL ATTAINMENT which represents	DECISION-MAKING and PLANNING
ADAPTATION which involves	RESOURCE MOBILIZATION and MANAGEMENT
INTEGRATION which comes from	COMMUNICATION and COORDINATION
PATTERN MAINTENANCE which requires	CONFLICT MANAGEMENT; it also involves "socialization," which goes beyond irrigation activities.

The prescriptive analysis compiled by Layton for water users' associations in Egypt (Sallam et al., 1984) proposed eight general "processes" which are drawn from the work of Haas and Drabek (1973), who were influenced by Parsons' writing. Their eight "processes," shown in the left-hand column below, are equivalent to our four but are more complicated and abstract as a set:

Processes (Haas and Drabek)	Activities
DECISION-MAKING, and TASK PROCESSES	DECISION-MAKING and PLANNING
ADAPTATION, and CONTROL PROCESSES	RESOURCE MOBILIZATION and MANAGEMENT
COMMUNICATION, and COORDINATION PROCESSES	COMMUNICATION and COORDINATION
CONFLICT, and MAINTENANCE PROCESSES	CONFLICT MANAGEMENT

Sallam et al. (1984) carry out their analysis with reference to four "periods" which parallel the four structurally-focused activities we have identified, though we see these more as possibly concurrent than as invariably sequential phases:

Periods (Sallam et al.)	Activities
PLANNING	DESIGN/PLANNING
ORGANIZATION	CONSTRUCTION/ IMPLEMENTATION
OPERATION	OPERATION
CONTINUATION	MAINTENANCE

This analysis suggests considerable convergence of thinking in the irrigation management area even if the designations used are sometimes different. In our analysis we have tried to use terminology that is common in everyday discourse, avoiding neologisms and jargon so as not to create new words or to attach unfamiliar meanings to familiar words.

Annex 3

Irrigation Management Roles

ASIA

MNC-TSC (Andhra Pradesh, India)

Elders (peddamanshula): operate as a committee, not compensated; represent villagers to officials, appoint common irrigators, sluice guards, and field guards; keep records; collect and manage the village fund; settle conflicts; determine planting, harvesting and other schedules

Common irrigators (neeruganti): serve 120 acres each, distributing water from channels to fields; implement rotation if water scarce; patrol canals; paid in grain by farmers

Sluice guards: patrol canals up to distributary outlet to guard against theft by upstream villages; up to 10 are employed during times of peak shortage, as a show of strength; paid by committee from common funds

Field guards: prevent damage by cattle or theft of grain; report violations to elders; serve irrigated and rainfed fields; paid monthly salary by committee

Sananeri (Tamil Nadu India)

President of Tank Committee: conducts meetings and oversees activities between meetings; manages finances; auctions off fishing rights in tank to raise funds for committee; pays for minor repairs; is consulted on the opening and closing of tank sluices; liaison between cultivators and Public Works Department, also with other tank committees served by same feeder channel

Water distributors (niirpaaycci): responsible for distributing water from channels to all fields; no maintenance duties (responsibility of farmers); paid in grain by farmers

Channel patrols (niiraaNi): those who bring the water down): responsible for maintaining authorized and adequate supply of water to tank; some maintenance of feeder channel and patrolling it to check theft or damage; coordinate with patrols from other tanks served by feeder

	channel; open and close tank sluices; inform irrigators of annual meeting; paid by tank committee
Chaj Doab (Punjab, Pakistan)	No specialized roles; "brotherhoods" (biradaris) provide framework for all social organization; large farmers coordinateirrigation activities if there is any coordination
Daudzai (N.W.F.P., Pakistan)	Elders (mashers): oversee community affairs including irrigation; carry out various irrigation tasks if there is no chawkidar in the community Village watchman-overseer (chawkidar): acts as ditch-tender for irrigation system; paid from community funds
Pul Eliya (Sri Lanka)	Irrigation headman (vel vidane): oversaw all irrigation activities; distributed water, organized maintenance, resolved disputes; paid in grain and also given some land to cultivate as compensation; role abolished 1958
Gal Oya (Sri Lanka)	Farmer-representative (govi-nivojitiya): selected by field channel group; oversees rotation and maintenance, encourages water saving and improved practices; represents farmers at higher levels in the system
Chhatis Mauja (Nepal)	Chairman and secretary of association: elected by assembly of cultivators; responsible, with other members of the executive committee, for operation of entire system; paid cash salary from funds contributed by members Technical supervisors (meth muktiya): employed by officers to oversee operation and maintenance of main system and advise on O&M at lower levels; often retired army personnel (gurkas); paid salary by association Messengers: employed by officers to communicate changes in operation schedule and to mobilize labor for maintenance; given small salary and use of bicycle
Argali and Chherlung (Nepal)	Headman (mukhiya): elected leader, mobilizes labor, directs work, excused from labor obligations

	Secretary: handles accounts, also excused from labor, may be paid if there is balance of funds remaining Patrols: patrol channels to guard system during monsoon season; do minor maintenance; paid for work
Tallo Kulo (Nepal)	Headman (*mukhiya*): head of organization, keeps accounts, mobilizes labor; hereditary position Elders (*kulo samiti*): committee of seven esteemed older persons who oversee irrigation work Patrols (*kulo poles*): guard against damage to system and against illicit tapping of water
Zanjeras (Philippines)	President (*cabecilla*), secretary (*papelista*), and treasurer (*tesoro*): usual duties for such positions; group may have two presidents, one internal and one external Cook: important role because feasts are significant part of social organization underlying irrigation cooperation Leaders of work groups (*gunglos*): mobilize and direct labor of small groups on maintenance and other tasks
Communal System (Philippines)	President, secretary and treasurer of Irrigators Service Association: usual duties for such positions Ditchtenders (*kanaleros*): traditional role incorporated into ISA structure for distributing water
Subaks (Indonesia)	Subak headman (*klian subak*): responsible for overseeing work and operation of the irrigation organization Tempak headman (*klian tempak*): responsible for work and operation of his irrigation group within organization Specialized work group (*pekaseh subak*): carry out most operation and maintenance activities on behalf of subak
Dharma Tirta (Indonesia)	Officers: formally elected set of officers, with usual responsibilities for overseeing irrigation Irrigation headman (*ulu-ulu*): traditional role,

coopted into Dharma Tirta, supervising O&M activities
Irrigation sub-headman (*pembantu ulu-ulu*): similar responsibilities as *ulu-ulu* for sub-group handling various O&M activities

Seraphi (Thailand)

Irrigation headman: role found at three levels -- village, major canal, and maximal canal; represents constituents' water needs to higher levels, transmits decisions, orders and information downward; resolves disputes, reports violations to district officials; organizes and supervises maintenance; keeps records of land farmed by water users; village-level headmen chosen by users, higher headmen by officials; excused from paying land taxes and providing labor; may keep some of the fines levied
Assistant village irrigation headmen: act as messengers between irrigation headman and farmers; coordinate irrigation activities of farmers in different sections of the village
Gatemen: guard the canal gates, open them and shut them when told to do so by irrigation officials

Sankamphaeng (Thailand)

Irrigation headman (*kae muang*): checks weirs and ditch conditions, supervises repairs, keeps accounts; chosen for fairness, honesty and technical competence, average period of service is 6 years; receives fines and gets an additional share of water
Messenger (*larm*): notifies members of meeting dates, brings materials for construction
Head of user association: new role, appointed by district administrator, with approval of water users; can be dismissed if majority of them are dissatisfied with him

Nam Tan (Laos)

Irrigation headman (*nai nam*): traditional role supervising users in construction and maintenance work; role taken over in government-sponsored system to handle water distribution and coordination with bureaucracy; paid in grain by users (16 kilograms of paddy per hectare)
Officers of user association: usual positions and responsibilities for irrigation management; 20

	farmer groups federated into Farmers Association with 900 members
Farm Land Improvement Association (South Korea)	Officers of FLIA: usual positions and responsibilities for irrigation management Canal patrollers: each to cover about 250 acres (up to 375 acres) on bicycle; patrol main and branch canals twice a day, adjust gates, check structures, read gauges, spot pest or disease attacks, give advice; must live in area he patrols; paid regular salary

MIDDLE EAST

Izki (Oman)	Agent (wakil): supervises operation of system; auctions off water shares to raise money for costs of system Water distributors (arifs): send water along channels to fields, expertise in minimizing transmission losses Secretary: guardian of the Falaj book; keeps records of water rights Technicians (awamir): builders of the qanats, do repairs Water witches (basir): diviners of water
Daghara (Iraq)	Tribal chief (shaykh): along with other duties as leader of the tribal ashira, oversees irrigation matters and represents water users to the authorities Honorables (sada): descendents of the Prophet Mohammad; supposed to encourage productive and amicable relations by exemplary life; resolve conflicts within community

AFRICA

Matam (Senegal)	Chairman: manages affairs of the pump irrigation group; places orders for inputs and distributes them, oversees pump operator

	Pump operator: trained 1-2 days, paid in-kind or in cash or given plot of land; remuneration left to group Bureau (executive committee): other officers in addition to the chairman -- vice-chairman, treasurer plus four members at large -- who act as intermediaries with members
Marakwet (Kenya)	Blowers: live in strategic locations where they will be the first to observe problems with canal system; sound an alarm through special horns to call members to make repairs
Mwea (Kenya)	Head cultivator: chosen by the operating authority's Field Assistant; given bicycle and excused from doing communal work; liaison between project staff and farmers
Gezira (Sudan)	Originally, wakil sheikhs were appointed by the project authorities to act as liaisons with "tenants"; later replaced by samads who filled similar role, but with no involvement in irrigation management either; now there are farmers as representatives on Tenant Representative Board

LATIN AMERICA

San Pedro de Atacama (Chile)	Junta de vigilencia de riego: elected officers plus seven delegates, one from each of the canal associations Treasurer: collects monthly fee from all members; informs Junta of delinquencies and may cut off water Ditchtender (celador): took place of Inspector de Aguas, which was too burdensome a role as it covered the whole area; inspector job was split into seven celador roles, one for each association; paid cash salary, makes twice daily rounds
Huerta Redonda (Ecuador)	President of association: resolves disputes that cannot be handled by water guards at field level; association is composed of those farmers who

took part in the battle with the neighboring hacienda in 1967 to get rights to water from the hacienda canal; the government finally assured them of use of the water when the hacienda was not using it (on week-ends) and the association sprung up immediately; the association is in effect embedded in the traditional communal organization (puebla) and does not as a rule hold separate meetings; occasional efforts to involve municipal authorities in resolving water disputes have not been satisfactory

Quinua (Peru)	Municipal authorities have been given responsibility for the Lurin Sayoc system, but they make decisions only about maintenance, not allocation or distribution, and there are many disputes and conflicts; farmers are still obliged to do maintenance work or pay a fine, but enforcement is uneven and so, as a result, is the work Irrigation judge: in Hanan Sayoc this traditional role has been maintained, to make weekly allocations of water and to assign maintenance responsibilities; fines are still collected from any farmers who do not do the work, so this system continues to operate reasonably well
Oaxaca (Mexico)	Office holders of groups (tramos) are not paid; they are linked to the traditional civil-religious hierarchy; titles vary from community to community as do the divisions of responsibility; roles are embedded in village government so office holders have non-irrigation responsibilities too

References

References

Abel, Martin E.
1975 Irrigation Systems in Taiwan: Management of a Decentralized Public Enterprise. Staff Paper ER 75-1. St. Paul: Department of Agricultural and Applied Economics, University of Minnesota.

Abeyratne, Shyamala, Jeffrey Brewer, Sena Ganewatte and Norman Uphoff
1984 Improving Irrigation Management Through Farmer Organization: Responses to a Program in Sri Lanka. Paper for SSRC-IIM/Bangalore seminar, January 4-7.

Abeyratne, Shyamala and Jayantha Perera
1984 Changes in Irrigation Management in Small Communities: A Tank and Anicut System in Moneragala District, Sri Lanka. Sri Lanka Journal of Agrarian Studies, 5(1), 71-84.

Abha, Sirovongs na Aytthaya
1979 A Comparative Study of Traditional Irrigation Systems in Two Communities in Northern Thailand. Bangkok: Social Research Institute, Chulalongkorn University.

Adams, Adrian
1977 The Senegal River Valley: What Kind of Change? Review of African Political Economy, 10(1), 33-59.

Adams, M. E.
1983 Water Management Extension among Small Farmers in Central Java. FAO Economic and Social Development Series No. 1. Rome: FAO.

Afifuddin bin Haji-Omar
1978 Peasants, Institutions and Development in Malaysia: The Political Economy of Development of the Muda Region. Ph.D. dissertation, Cornell University.

ADB
1983 Operational Manual for the Irrigated Agriculture in the Nong Wai Pioneer Agriculture Project. Kingdom of Thailand, Ministry of Agriculture and Cooperatives. Manila: Asian Development Bank.

Bagadion, Benjamin U. and Frances F. Korten
1985 Developing Irrigators' Organizations: A Learning Process Approach to a Participatory Irrigation Program. In Cernea (1985:52-90).

Bailey, Charles, Robert Chambers and Roberto Lenton
1981 Final Report: Water Management Consultancy: Egypt and Sudan. New Delhi: Ford Foundation.

Baldwin, K. D. S.
1957 The Niger Agricultural Project: An Experiment in African Development. Oxford: Basil Blackwell.

Benedict, Peter et al.
1981 Sudan: The Rahad Irrigation Project. AID Project Impact Evaluation Report No. 31. Washington: USAID.

Bhatty, K. M.
1979 Social Determinants of Water Management in Daudzai. Peshawar: Pakistan Academy for Rural Development.

Birkelbach, Aubrey
1973 The Subak Association. Indonesia, 153-169. Ithaca: Southeast Asia Program, Cornell University.

Bottrall, Anthony W.
1981 Comparative Study of the Management and Organization of Irrigation Projects. Staff Working Paper No. 458. Washington: World Bank.

1981a Improving Canal Management: The Role of Evaluation and Action Research. Water Supply and Management, 5(1), 67-79.

1985 Managing Large Irrigation Schemes: A Problem of Political Economy. Agricultural Administration Unit Occasional Paper No. 5. London: Overseas Development Institute.

Bromley, Daniel
1982 Improving Irrigated Agriculture: Institutional Reform and the Small Farmer. Staff Working Paper No. 531. Washington: World Bank.

Cabanilla, L. S.
1985 Study of Operation and Maintenance Problems in Irrigation: The Philippine Case. Report for USAID. Manila: USAID.

Calavan, Michael
1984 Appropriate Administration: Creating a "Space" Where Local Initiative and Voluntarism Can Grow. In Private Voluntary Organizations as Agents of Change, ed. Robert F. Gorman, 211-249. Boulder: Westview Press.

Cernea, Michael
1983 A Social Methodology for Community Participation in Local Investments: The Experience of Mexico's PIDER Program. Staff Working Paper No. 598. Washington: World Bank.

1984 The Social Organization of Water Users: Experience with Water Users Associations in Bank Projects. Draft paper. Washington: World Bank.

1984a Can Local Participation Help Development? Mexico's PIDER Program Shows It Can Improve the Selection and Execution of Local Development Projects. Finance and Development, 21(4), 41-44.

1985 Putting People First: Sociological Variables in Rural Development, editor. New York: Oxford University Press.

Chambas, Mohamed Ibn
1980 The Politics of Agricultural and Rural Development in the Upper Region of Ghana: Implications of Technocratic Ideology and Non-Participatory Development. Ph.D. dissertation, Cornell University.

Chambers, Robert
1975 Water Management and Paddy Production in the Dry Zone of Sri Lanka. Colombo: Agrarian Research and Training Institute.

1977 Men and Water: The Organization and Operation of Irrigation. In Green Revolution?: Technology and Change in Rice-Growing Areas of Tamil Nadu and Sri Lanka, ed. B. H. Farmer, 340-363. Boulder: Westview Press. Reprinted in Coward (1980:28-50).

1980 In Search of a Water Revolution: Questions for Managing Canal Irrigation in the 1980s. In Irrigation Water Management, 23-37. Los Banos: International Rice Research Institute.

1983 Rural Development: Putting the Last First. London: Longman.

1984 Farmers Above the Outlet: Irrigators and Canal Management in South Asia. Draft paper. Sussex: Institute of Development Studies.

Chambers, Robert and Jon Moris, eds.
1973 Mwea: An Irrigated Rice Settlement Scheme in Kenya. Munich: Weltforum Verlag.

Cloud, Kathleen
1982 Women in Water Management: Understandings Gained from My Asian Trip. Report to Water Management Synthesis II Project, November 30, 1982. Tucson: Consortium for International Development, mimeo.

Cohen, John M. and Norman Uphoff
1977 Rural Development Participation: Concepts and Measures for Project Design, Implementation and Evaluation. Ithaca: Rural Development Committee, Cornell University.

CSU/CID
1980 Problem Identification Report for the Kafr el Sheikh Study Area. Technical Report No. 6. Cairo: Water Distribution Research Institute, Water Research Center, Ministry of Irrigation.

Corey, Gilbert L. and Wayne Clyma
1975 Improving Farm Water Management in Pakistan. Water Management Technical Report No. 37. Fort Collins: Colorado State University.

Cornick, Tully
1983 The Social Organization of Production in Quimiag, Ecuador: A Case Study of Small Farmer Production Systems in the Highland Andes. Ph.D. dissertation, Cornell University.

Coward, E. Walter, Jr.
1971 Agrarian Modernization and Village Leadership: Irrigation Leaders in Laos. Asian Forum, 3, 158-163.

1976 Indigenous Organization, Bureaucracy and Development: The Case of Irrigation. Journal of Development Studies, 13, 92-105. Reprinted in Coward (1980:329-399).

1977 Irrigation Management Alternatives: Themes from Indigenous Irrigation Systems. Agricultural Administration, 4, 233-257.

1979 Principles of Social Organization in an Indigenous Irrigation System. Human Organization, 38(1), 28-36.

1980 Irrigation and Agricultural Development in Asia: Perspectives from the Social Sciences, ed. Ithaca: Cornell University Press.

1980a Irrigation Development: Institutional and Organizational Issues. In Coward (1980:15-27).

1983 Property in Action: Alternatives for Irrigation Investment. Paper for Workshop on Water Management and Policy, Khon Kaen University, Khon Kaen, Thailand.

1984 Improving Policies and Programs for the Development of Small-Scale Irrigation Systems. WMS Report No. 27. Ft. Collins: Water Management Synthesis Project, Colorado State University.

1985 Traditional Irrigation Systems and Government Assistance: Current Research Findings from Southeast Asia. DVWK Bulletin, 9(1), 1-30.

1985a Technical and Social Change in Currently Irrigated Regions: Rules, Roles and Rehabilitation. In Cernea (1985:27-51).

Coward, E. Walter and Norman Uphoff

1986 Operation and Maintenance in Asian Irrigation: Reappraising Government and Farmer Responsibilities and Rights. Irrigation and Drainage Systems, forthcoming.

de los Reyes, Romana P.

1984 Sociotechnical Profile: A Tool for Rapid Rural Appraisal. Quezon City: Ateneo de Manila University.

de Silva, N. G. R.

1981 Farmer Participation in Water Management: The Minipe Project in Sri Lanka. Rural Development Participation Review, 3(1), 16-19.

1984 Involvement of Farmers in Water Management: Alternative Approach at Minipe, Sri Lanka. In FAO (1985:133-148).

de Wilde, John

1967 Mali -- The Office du Niger: An Experiment with Irrigated Agriculture. In J. de Wilde et al., Experiences with Agricultural Development in Tropical Africa, Vol. III: The Case Studies. Baltimore: Johns Hopkins University Press.

Diemer, Geert and Ellen van der Laan
1983 Small Scale Irrigation Along the Senegal River. Irrigation Management Network Paper 8f. London: Overseas Development Institute.

Doan, Peter, David Douglin and Norman Uphoff
1984 Local Institutional Development for Natural Resource Management. Ithaca: Rural Development Committee, Cornell University. Revised version in Uphoff (1986a: Chapter 2).

Doan, Rebecca Miles, Gregory Schmidt and Norman Uphoff
1984 Mobilizing and Managing Economic Resources for Local Institutional Development. Ithaca: Rural Development Committee, Cornell University. Revised version in Uphoff (1986a: Chapter 8).

Downing, Theodore
1974 Irrigation and Moisture-Sensitive Periods: A Zapotec Case. In Downing and Gibson (1974:113-122).

Downing, Theodore and McGuire Gibson, eds.
1974 Irrigation's Impact on Society. Anthropological Papers of the University of Arizona No. 25. Tucson: University of Arizona Press.

Duewel, John
1982 Promoting Participatory Approaches to Cultivating Water Users Associations: Two Case Studies from Central Java. Technical Report No. 11. Ithaca: Departments of Rural Sociology and Agricultural Engineering, Cornell University.

1984 Central Java's Dharma Tirta WUA "Model": Peasant Irrigation Organisations under Conditions of Population Pressure. Agricultural Administration, 17(4), 261-285.

Edmonds, G.A.
1980 The "Roads and Labour" Programme, Mexico. In G.A. Edmunds and J.D.F.G. Howe, eds., Roads and Resources: Appropriate Technology in Road Construction in Developing Countries, 123-134. London: Intermediate Technology Development Group, for ILO.

Eggink, Jan Willem and Jan Ubels
1984 Irrigation, Peasants and Development. Wageningen: Department of Irrigation and Civil Engineering and Department of Rural Sociology, Agricultural University of Wageningen.

Elder, John
1982 Problems in Sudan's Gezira Irrigation Scheme. Unpublished paper, Department of Rural Sociology, Cornell University.

Esman, Milton J. and Norman Uphoff
1984 Local Organizations: Intermediaries in Rural Development. Ithaca: Cornell University Press.

Fairchild, Warren
1985 Institutional Aspects: Water Users Associations (Pakistan). Paper for Second Irrigation and Drainage Seminar, World Bank, December 17-19.

FAO
1982 Farmers Participation and Organization for Irrigation Water Management. Rome: International Support Programme for Farm Water Management, FAO.

1985 Participation Experiences in Irrigation Water Management. Proceedings of the Expert Consultation on Irrigation Water Management, Yogyakarta and Bali, July 16-22, 1984. Rome: FAO.

Fernea, Robert A.
1970 Shaykh and Effendi: Changing Patterns of Authority among the El Shabana of Southern Iraq. Cambridge: Harvard University Press.

Fleuret, Patrick
1985 The Social Organization of Water Control in the Taita Hills, Kenya. American Ethnologist, 32(1), 103-118.

Freeman, David M. and Max L. Lowdermilk
1985 Middle-level Organizational Linkages in Irrigation Projects. In Cernea (1985: 91-118).

Fresson, Sylvia
1979 Public Participation in Village-Level Irrigation Perimeters in the Matam Region of Senegal. In Self-Help and Popular Participation in Rural Water Systems, ed. Duncan Miller, 96-137. Paris: Development Center, OECD.

Gaitskell, A.
1959 Gezira: A Story of Development in the Sudan. London: Faber and Faber.

Geertz, Clifford
1967 Organization of the Balinese Subak. In Villages in Indonesia, ed. Koentjaraningrat. Ithaca: Cornell University Press. Excerpted in Coward (1980:70-90).

Gray, Robert F.
1963 The Sonjo of Tanganyika: An Anthropological Study of an Irrigation-Based Society. London: Oxford University Press.

Haas, J. Eugene and Thomas Drabek
1973 Complex Organizations: A Sociological Perspective. New York: Macmillan.

Hall, Anthony L.
1978 Drought and Irrigation in North-East Brazil. Cambridge: Cambridge University Press.

Hart, Henry C.
1978 Anarchy, Paternalism or Collective Responsibility Under the Canals? Economic and Political Weekly, 13(39), A125-134.

Hashim Ali, Syed et al.
1982 Report of the Commission for Irrigation Utilisation. Hyderabad: Government of Andhra Pradesh.

Howes, Michael
1984 The Social Organization of Deep Tubewell Irrigation in Bangladesh. Paper for SSRC-IIM/Bangalore seminar, January 4-7.

Hunt, Robert and Eva C. Hunt
1974 Irrigation, Conflict, and Politics: A Mexican Case. In Downing and Gibson (1974: 129-157).

1976 Canal Irrigation and Local Social Organization. Current Anthropology, 17(3), 389-411.

Hutupea, R., P. Dirjasanyata and N. G. S. Nordoit
1979 The Organization of Farm-Level Irrigation in Indonesia. In Irrigation Policy and the Management of Irrigation Systems in Southeast Asia. Bangkok: Agricultural Development Council.

Illo, Jean Frances and Ma. Elena Chiong-Javier
1983 Organizing Farmers for Irrigation Management: The Buhi-Lalo Experience. Naga City: Research and Service Center, Ateneo de Naga.

Illo, Jean Frances and Felix Nestor
1981 Organizing Farmers for Irrigation Management: The Development of a New System, Lower Lalo National Irrigation Project. Naga City: Research and Service Center, Ateneo de Naga.

ISTI
1985 Final Evaluation of Sri Lanka Water Management (Project No. 383-0057). Washington: International Science and Technology Institute, Inc.

Inos, Faustino I.
1981 A Survey of Farmers' Participation in the Development, Operation and Management of Selected Irrigation Systems in the Philippines. Ph.D. dissertation, Rutgers University.

Jayaraman, T. K.
1981 Farmers' Organizations in Surface Irrigated Projects: Two Empirical Studies from Gujarat. Economic and Political Weekly, 16(39), A113-116.

1984 Socio-Economic Determinants of Farmer Participation in Irrigation Projects in India. Journal of Rural Development, 7(1), 77-100. Seoul: Korean Rural Economics Institute.

Johnson, Sam H., III
1982 Large-Scale Irrigation and Drainage Schemes in Pakistan: A Study of Rigidities in Public Decision-Making. Food Research Institute Studies, 18(2), 173-180.

Kahn, Miriam
1983 In Fear of an Aqueduct: Irrigation Systems as Potentially Disruptive Forces. Paper for American Anthropological Association meetings, Chicago.

Kathpalia, G. N.
1984 The Nong Wai Irrigation Project, Thailand. Irrigation Management Network Paper 10c. London: Overseas Development Institute.

Kelly, William W.
1983 Irrigation Management in Japan: A Critical Review of Japanese Social Science Research. Ithaca: Rural Development Committee, Cornell University.

Kikuchi, Masao, Geronimo Dozina, Jr. and Yujiro Hayami
1978 Economics of Community Works Programs: A Communal Irrigation Project in the Philippines. Economic Development and Cultural Change, 26(2), 211-225.

Korten, David C.
1980 Community Organization and Rural Development: A Learning Process Approach. Public Administration Review, 40(5), 480-511.

1982 The Working Group as a Mechanism for Managing Bureaucratic Reorientation: Experience from the Philippines. NASPAA Working Paper No. 4. Washington: National Association of Schools of Public Affairs and Administration.

Korten, David C. and Norman Uphoff
1981 Bureaucratic Reorientation for Participatory Rural Development. NASPAA Working Paper No. 1. Washington: National Association of Schools of Public Affairs and Administration.

Korten, Frances F.
1982 Building National Capacity to Develop Water Users' Associations: Experience from the Philippines. Staff Working Paper No. 528. Washington: World Bank.

Kottak, Conrad Phillip
1985 When People Don't Come First: Some Sociological Lessons from Completed Projects. In Cernea (1985:325-356).

Lamb, Geoff and Linda Mueller
1982 Control, Accountability and Incentives in a Successful Development Institution: The Kenya Tea Development Authority. Staff Working Paper No. 550. Washington: World Bank.

Laucoin, Guy
1971 Vers l'Autogestion des Amenagements Hydro-agricole au Niger. Developpment et Civilisations, 44(7), 54-62.

Leach, E. R.
1961 Pul Eliya: A Village in Ceylon. Cambridge: Cambridge University Press. Excerpted in Coward (1980:91-126).

Lees, Susan
1973 Socio-Political Aspects of Canal Irrigation in Oaxaca. Memoirs of the Museum of Anthropology No. 6. Ann Arbor: University of Michigan.

Levine, Gilbert and Henry C. Hart
1981 Mobilizing Local Resource for Irrigation. Report No. 22. New York: Agricultural Development Council.

Lewis, Henry T.
1971 Ilocano Rice Farmers: A Comparative Study of Two Philippine Barrios. Honolulu: University of Hawaii Press. Excerpted in Coward (1980:153-171).

Lowdermilk, Max
1985 Improved Irrigation Management: Why Involve Farmers? Irrigation Management Network Paper 11c. London: Overseas Development Institute.

Lowdermilk, Max, Wayne Clyma and Alan Early
1975 Physical and Socio-Economic Dynamics of a Watercourse in Pakistan's Punjab: System Constraints and Farmer Responses. Water Management Technical Report No. 42. Fort Collins: Colorado State University.

Lynch, Barbara
1978 A Desert Agricultural System in a Mineral Export Enclave: Water Management and Productive Capacity in San Pedro de Atacama. M.S. thesis, Cornell University.

1983 Trip Report: Peru and Bolivia. Ithaca: Water Management Synthesis Project, Cornell University.

1985 Community Participation and Local Organization for Small-Scale Irrigation. WMS Report No. 34. Ithaca: Water Management Synthesis Project, Cornell University.

Maass, Arthur and Raymond L. Anderson
1978 ...and the Desert Shall Rejoice: Conflict, Growth and Justice in Arid Environments. Cambridge: Massachusetts Institute of Technology Press.

Martin, Edward D.
1986 Resource Mobilization, Water Allocation and Farmer Organization in Hill Irrigation Systems in Nepal. Ph.D. dissertation, Cornell University.

Martin, Edward and Robert Yoder
1983 Water Allocation and Resource Management for Irrigation: A Comparison of Two Systems in Nepal. Paper for Nepal Studies Association, Madison, November.

Mayson, Evan
1984 Assisting Villagers to Build Their Own Irrigation Scheme: An Example from Thailand. Irrigation Management Network Paper 9e. London: Overseas Development Institute.

Meinzen-Dick, Ruth S.
1984 Local Management of Tank Irrigation in South India: Organization and Operation. Cornell Studies in Irrigation No. 3. Ithaca: Cornell University.

Merrey, Douglas J.
1982 Reorganizing Irrigation: Local Level Management in the Punjab (Pakistan). In Desertification and Development: Dryland Ecology in Social Perspective, eds. H. Mann and B. Spooner, 83-109. London: Academic Press.

1983 Irrigation, Poverty and Social Change in a Village of Pakistani Punjab: An Historical and Cultural Ecological Analysis. Ph.D. dissertation, University of Pennsylvania.

1983a Irrigation and Underdevelopment in Punjab, Pakistan. Paper for meetings of American Anthropological Association, November.

1984 Canal Irrigation, Colonialism and Underdevelopment: The Roots of Poverty in an Irrigated Village in Punjab, Pakistan. Paper for SSRC-IIM/Bangalore seminar, January 4-7.

Mitchell, William P.
1976 Irrigation and Community in the Central Peruvian Highlands. American Anthropologist, 78(7), 24-44.

Montgomery, John D.
1983 When Local Participation Helps. Journal of Policy Analysis and Management, 3(1), 90-105.

Moore, M. P.
1979 Social Structure and Institutional Performance: Local Farmers' Organizations in Sri Lanka. Journal of Administration Overseas, 18(4), 240-249.

1980 Approaches to Improving Water Management in Large-Scale Irrigation Schemes in Sri Lanka. Occasional Paper No. 20. Colombo: Agrarian Research and Training Institute.

1983 Irrigation Management in Taiwan. Unpublished paper. Sussex: Institute of Development Studies.

1984 Institutional Development, The World Bank and India's New Agricultural Extension Programme. Journal of Development Studies, 20(4), 303-317.

Moore, M. P., F. Abeyratne, R. Amarakoon, J. Farrington
1983 Space and the Generation of Socio-Economic Inequality on Sri Lanka's Irrigation Schemes. MARGA, Quarterly Journal of the MARGA Institute, Colombo, 7(1), 1-32.

Morfit, Michael
1983 Community Management of Small Scale Irrigation: The Case of the High Performance Sedarhana Irrigation System Pilot Project in Indonesia. Jakarta: USAID.

Moris, John R. and Derrick J. Thom
1985 African Irrigation: An Overview; Summary Report. WMS Report No. 37. Logan: Water Management Synthesis Project, Utah State University.

Murray-Rust, D. Hammond and M. P. Moore
1984 Formal and Informal Water Management Systems: Cultivation Meetings and Water Delivery in Two Sri Lankan Irrigation Schemes. Cornell Studies in Irrigation No. 2. Ithaca: Cornell University.

Nickum, James, ed.
1981 Management of the Meichuan Reservoir Irrigation Project. Water Management Organization in the People's Republic of China. Armonk, New York: M. E. Sharpe Publishers.

Palanasami, K. and William Easter
1983 The Tanks of South India: A Potential for Future Expansion and Irrigation. Staff Paper ER 83-4. St. Paul: Department of Agricultural and Applied Economics, University of Minnesota.

Pant, Niranjan
1981 Utilization of Canal Water Below Outlet in Kosi Irrigation Project: Administrative and Community Level Solutions. Economic and Political Weekly, 16(39), A78-88.

1984 Productivity and Equity in Irrigation Systems, editor. New Delhi: Ashish Publishing House.

Pant, Niranjan and R.K. Verma
1983 Farmers' Organization and Irrigation Management. New Delhi: Ashish Publishing House.

Patil, R. K. and D. N. Kulkarni
1984 Ensuring Equity in Irrigation Systems: A Case Study of a Farmer Group-Managed System. In Pant (1984:199-208).

Patterson, William
1984 Village Irrigation Systems Along the Senegal River: A Comparative Analysis. Ph.D. dissertation, Cornell University.

Plusquellec, Herve L. and Thomas Wickham
1985 Irrigation Design and Management: Experience in Thailand and Its General Applicability. Technical Paper No. 40. Washington: World Bank.

Potter, Jack M.
1976 Thai Peasant Social Structure. Chicago: University of Chicago Press.

Pradhan, Prachanda P.
1983 Chhatis Mauja. In Water Management in Nepal: Proceedings of a Seminar on Water Management Issues. Kathmandu: Agricultural Projects Service Centre.

1984 Chhatis Mauja Irrigation System: Community Response in Resource Management. Paper for SSRC-IIM/Bangalore seminar, January 4-7.

Pradhan, Ujjwal
1982 Irrigation Development: Whose Panacea? Mimeographed paper.

Radosevich, George
1975 Water User Organizations for Improving Irrigated Agriculture: Applicability to Pakistan. Fort Collins: Water Management Research Project, Colorado State University.

Ramachandran, V. K.

1984 Irrigated Agriculture and Agricultural Labor: A Study of the Cumbum Valley, Madurai District. Paper for SSRC/IIM-Bangalore seminar, January 4-7.

Rao, P. K.

1984 Comments on Cost Recovery and Irrigation Water Pricing. Irrigation Management Network Paper 10f. London: Overseas Development Institute.

Reidinger, Richard

1974 Institutional Rationing of Canal Water in Northern India: Conflict between Traditional Patterns and Modern Needs. Economic Development and Cultural Change, 23(1), 79-104. Reprinted in Coward (1980:263-288).

Robinson, David M.

1982 Water Users' Organizations in Two Large Philippine Irrigation Systems: Constraints and Benefits of a Participatory Approach to Water Management. Ph.D. dissertation, Cornell University.

1985 Constraints on Farmer Participation in Government-Supported Irrigation Systems in Indonesia. Draft paper. Jakarta: USAID.

Roder, Wolf

1965 The Sabi Valley Irrigation Projects. Department of Geography Research Paper No. 99. Chicago: University of Chicago.

Rondinelli, Dennis

1983 Development Projects as Policy Experiments: An Adaptive Approach to Development Administration. New York: Methuen.

Sallam, Mohamed et al.

1984 Experience with Water Users' Associations, as compiled by James J. Layton. Egypt Water Use and Management Project, Technical Report No. 65. Cairo: Water Distribution Research Institute, Water Research Center, Ministry of Irrigation.

Sasaki, Takashi
1985 The Role of Village Collective Management Traditions in the Formation of Group Farming: Case Studies in Japan. Regional Development Dialogue, 6(1), 66-103.

Seckler, David
1985 The Management of Padi Irrigation Systems: A Laissez-Faire Supply-Side Theory. Irrigation Management Network Paper 11b. London: Overseas Development Institute.

Sengupta, Nirmal
1980 The Indigenous Irrigation Organization in South Bihar. Indian Economic and Social History Review, 27(2), 157-189.

1984 Ahar-Pyne Irrigation. Paper for SSRC-IIM/Bangalore seminar, January 4-7.

Shrestha, Bihari K.
1980 Nuwakot District, Nepal. In The Practice of Local-Level Planning: Case Studies in Selected Rural Areas of India, Nepal and Malaysia. Bangkok: U. N. Economic and Social Commission for Asia and Pacific.

Simpson, I. G.
1976 The Sudan Gezira Scheme: Transformation and Evolution. Paper for IADS Seminar on Accelerating National Agricultural and Rural Development, University of Reading.

Sims, Holly
1986 The State and Rural Development: Agricultural Policy and Performance in the Indian and Pakistani Punjabs. Book manuscript, Department of Political Science, University of California, Berkeley.

Singh, K. K.
1982 Farmers' Participation in Irrigation Management: The Pochampad Experience. Varanasi: Gandhian Institute of Studies.

1983 Assistance Needs of Water User Associations in Their First Years: The Example of Pochampad. Irrigation Management Network Paper 8a. London: Overseas Development Institute.

1984 Farmers' Associations in Large Irrigation Projects: The Pochampad Experience. Paper for SSRC-IIM/Bangalore seminar, January 4-7.

Siy, Robert Y.
1982 Community Resource Management: Lessons from the Zanjera. Quezon City: University of the Philippines Press.

Spooner, Brian
1974 Irrigation and Society: The Iranian Plateau. In Downing and Gibson (43-57).

Ssennyonga, Joseph
1983 The Marakwet Irrigation System as a Model of a Systems Approach to Water Management. In Kerio Valley: Past, Present and Future, eds. B. E. Kipkorir, R. Soper and J. Ssennyonga. Nairobi: Eleza Services Ltd.

Stavis, Benedict
1982 Rural Local Governance and Agricultural Development in Taiwan. In Rural Development and Local Organization in Asia, ed. N. Uphoff, Volume II: East Asia, 165-271. New Delhi: Macmillan.

Steinberg, David, Robert B. Morrow and Kim Dong-Il
1980 Korean Irrigation. AID Project Impact Evaluation Report No. 12. Washington: USAID.

Steinberg, David et al.
1983 Irrigation and AID Experience: Considerations Based on Evaluation. AID Project Evaluation Report No. 8. Washington: USAID.

Sundar, A.
1985 Managing Main System Water Distribution. WAMANA: A Quarterly Newsletter on Water Management, 5(4), 1-22. Bangalore: Irrigation Research Academy.

Sutton, Sally
1984 The Falaj: A Traditional Cooperative System of Water Management. Waterlines: Journal of Appropriate Water Supply and Sanitation Technologies, 2(3), 8-12.

Svendsen, Mark
1983 Water Management Strategies and Practices at the Tertiary Level: Three Philippine Irrigation Systems. Ph.D. dissertation, Cornell University.

Taylor, Donald C. and Effendi Pasandaran
1979 'Pasten': A Key Water Management Concept in Eastern Javanese Irrigation. In Irrigation Policy and the Management of Irrigation Systems in Southeast Asia. Bangkok: Agricultural Development Council.

Tiffen, Mary
1983 Alternative Design Concepts for Irrigation Systems: Definitions and Objectives. Irrigation Management Network Paper 8e. London: Overseas Development Institute.

Uphoff, Norman
1982 Contrasting Approaches to Water Management Development in Sri Lanka. In Third World Legal Studies: 1982, 202-249. New York: International Center for Law and Development.

1984 Analyzing Options for Local Institutional Development. Ithaca: Rural Development Committee, Cornell University. Revised version in Uphoff (1986a: Chapter 7).

1985 People's Participation in Water Management: Gal Oya, Sri Lanka. In Public Participation in Development Planning, ed. J. C. Garcia-Zamor, 131-178. Boulder: Westview Press.

1985a Developing Water Management Institutions in Sri Lanka: Understanding Collective Action and Managing Common Resources. Paper for American Society for Public Administration meetings, Indianapolis, March.

1985b Fitting Projects to People. In Cernea (1985:359-395).

1986 Activating Community Capacity for Water Management: Experience from Gal Oya, Sri Lanka. In Community Management: Asian Experience and Perspectives, ed. David C. Korten. West Hartford: Kumarian Press, forthcoming.

1986a Local Institutional Development: An Analytical Sourcebook with Cases. West Hartford: Kumarian Press.

Uphoff, Norman and Milton J. Esman

1974 Local Organization for Rural Development: Analysis of Asian Experience. Ithaca: Rural Development Committee, Cornell University.

Uphoff, Norman, Ruth Meinzen-Dick and Nancy St. Julien

1985 Improving Policies and Programs for Farmer Organization and Participation in Irrigation Water Management. Professional Paper No. 1. Ithaca: Water Management Synthesis Project, Cornell University.

Uphoff, Norman and Katy Van Dusen

1984 Local Institutional Development for Agriculture. Ithaca: Rural Development Committee, Cornell University. Revised version in Uphoff (1986a: Chapter 5).

Uphoff, Norman and R. D. Wanigratne

1982 Rural Development and Local Organization in Sri Lanka. In Rural Development and Local Organization in Asia, ed. N. Uphoff, Volume I: South Asia, 479-549. New Delhi: Macmillan.

Uphoff, Norman, M. L. Wickramasinghe and C. M. Wijayaratna

1981 'Optimum' Participation in Water Management: Issues and Evidence from Sri Lanka. Paper for Rural Development Committee, Cornell University, and Agrarian Research Training Institute, Colombo.

USAID
1984 A.I.D. Policy Paper: Local Organizations in Development. Washington: Bureau of Program and Policy Coordination, USAID.

Valera, Alfredo B.
1985 A Comparative Assessment of Three Irrigation Systems in Central Luzon, Philippines. Ph.D. dissertation, Cornell University.

Vander Velde, Edward J.
1980 Local Consequences of a Large-Scale Irrigation System in India. In Coward (1980:299-328).

Visaya, Benito P.
1982 The Palsiguan Multi-Purpose Project and the Zanjeras. Paper for Conference on Organization as a Strategic Resource in Irrigation Development, Asian Institute of Management, Manila, November 15-19.

Wade, Robert
1979 The Social Response to Irrigation: An Indian Case Study. Journal of Development Studies, 16(1), 3-26.

1980 Access to the Irrigation Department: The Tail-End Problem Under South Indian Irrigation Canals. Indian Journal of Public Administration, 26(2), 359-377.

1981 Water User Associations: Social Principles and Governmental Process. Unpublished paper cited in Montgomery (1983).

1982 Group Action for Irrigation. Economic and Political Weekly, 17(39), A103-106.

1982a The System of Administrative and Political Corruption: Canal Irrigation in South India. Journal of Development Studies, 18(3), 287-328.

1982b Irrigation and Agricultural Politics in South Korea. Boulder: Westview Press.

1984 The Market for Public Office: Why the Indian State Is Not Better at Development. Discussion Paper 194. Sussex: Institute of Development Studies.

1984a The Community Response to Irrigation: Management of the Water Commons in Upland South India. Paper for SSRC/IIM-Bangalore seminar, January 4-7.

Wade, Robert and Robert Chambers

1980 Managing the Main System: Canal Irrigation's Blind Spot. Economic and Political Weekly, 15(39), A107-112.

Weeramunda, A.J.

1985 Water Management in Kimbulwana Oya: A Case Study of Technical Leadership and Farmer Participation. Colombo: Agrarian Research and Training Institute, draft.

Whyte, William F. and Damon Boynton

1983 Higher Yielding Human Systems for Agriculture. Ithaca: Cornell University Press.

Widanapathirana, A. S.

1984 The Gal Oya Experiment. Irrigation Management Network Paper 10c. London: Overseas Development Institute.

Wijayaratna, C. M.

1984 Involvement of Farmers in Water Management in Gal Oya, Sri Lanka. In FAO (1985:117-132).

1986 Assessing Irrigation System Performance: A Methodological Study with Application to Gal Oya, Sri Lanka. Ph.D. dissertation, Cornell University.

Wilkinson, J. C.

1977 Water and Tribal Settlement in Southeast Arabia: A Study of the Aflaj of Oman. Oxford: Clarendon Press.

Wood, Geoffrey D.

1984 Provision of Irrigation Services by the Landless: An Approach to Agrarian Reform in Bangladesh. Agricultural Administration, 17(1), 55-80.

Yoder, Robert

1986 The Performance of Farmer Managed Irrigation Systems in the Hills of Nepal. Ph.D. dissertation, Cornell University.

Author Index

Abeyratne, 81
Abha, 50, 56
Adams, A., 61, 119
Adams, M., 56
Afifuddin, 24, 65, 69
Anderson, 133, 141

Bagadion, 28, 111, 150
Baldwin, 25
Benedict, 91
Bhatty, 51
Birkelbach, 72
Bottrall, ix, 3, 7, 8

Calavan, 124
Cernea, 77, 87, 116, 121, 123, 124, 126, 127, 129, 140, 142, 143, 155
Chambers, 5, 7, 13, 27, 30, 43, 76, 77, 84, 126, 142, 146, 149, 166
Chiong-Javier, 24, 87, 137, 147
Cloud, 92
Clyma, 57
Cohen, 13
Coward, 6, 13, 23, 62, 69, 73, 82, 87, 95, 117, 121, 126, 132, 165
CSU (Colorado State University), 120

de los Reyes, 139
de Silva, 34, 57, 74, 150, 156
Doan, P., 129
Doan, R., 56
Downing, 44, 48, 52, 69, 73, 78, 85, 99
Drabek, 167
Duewel, 44, 49, 73, 87, 132

Early, 57
Easter, 91
Egginck, 54
Esman, 13, 113, 124, 129, 133

Fairchild, 25, 124
Fernea, 57, 65, 103
Fleuret, 46, 116
Freeman, 165
Fresson, 50, 61

Geertz, 44, 72, 77, 90, 106, 132
Goldsmith, 13
Gray, 90, 125
Gunston, 105

Haas, 167
Hall, 31, 120
Hart, 146

Hashim Ali, 140
Howes, 61
Hunt, E., 54, 81, 165
Hunt, R. 54, 81, 165
Hutupea, 132

Illo, 24, 87, 137, 147
ISTI, 33

Johnson, 55

Kahn, 103
Kathpalia, 135, 141, 144, 150
Kelly, 165
Korten, D., 30, 112, 113, 120, 121, 123, 138, 153
Korten, F., 23, 28, 30, 111, 123, 132, 141, 150, 153
Kottak, 124
Kulkarni, 77

Lamb, 156
Laucoin, 62
Leach, 44, 52, 56, 100, 102
Lees, 11, 50, 52, 78
Lewis, 72
Lowdermilk, 24, 44, 51, 57, 138, 140, 165
Lynch, 31, 82, 124-125, 132, 139, 141, 155

Maass, 133, 141
Martin, 12, 25, 55, 84, 105, 125
Mayson, 137, 154
Meinzen-Dick, 54, 64, 90, 102, 135
Merrey, 31, 51, 65, 77, 85, 100, 102, 103
Mitchell, 44, 45, 48, 55, 88, 125
Montgomery, 110, 123
Moore, 45, 52, 62, 77, 83, 90, 140
Morfit, 27, 117
Moris, 27, 76, 95
Mueller, 156

Palanasami, 91
Pant, 147
Patil, 77
Perera, 81
Plusquellec, 44, 141
Pradhan, P., 46, 106
Pradhan, U., 12, 119

Radosevich, 124
Ramachandran, 90, 102
Rao, 156
Reidinger, 55
Robinson, 27, 32-33, 44, 144
Rondinelli, 113

Sallam, 167
Sasaki, 91
Seckler, 105
Sengupta, 69
Shrestha, 121
Sims, 55
Singh, 31, 49, 72, 87, 131, 151
Siy, 44, 69, 72, 74, 125
Spooner, 32, 47, 82, 98
Ssennyonga, 32
Steinberg, 140
Sundar, 5
Sutton, 47, 98
Svendsen, 23, 99, 121

Thom, 95
Tiffen, 161

Ubels, 54
Uphoff, 13, 23, 31, 56, 57, 84, 85, 89, 90, 104, 106, 113, 120, 124, 129, 130, 133, 142, 143, 150
USAID, 124, 127

Valera, 83
Vander Velde, 55
Visaya, 116
Van Dusen, 85

Wade, 5, 51, 54, 69, 76, 84, 85, 88, 90, 91, 106, 120, 126, 132, 146, 156
Wanigaratne, 57
Weeramunda, 150
Wickham, 44, 141
Widanapathirana, 102, 150
Wijayaratna, 29, 33, 126, 150
Wilkinson, 32, 47, 91, 98, 103
Wood, 142

Yoder, 12, 25, 44, 44, 105, 125

Country/Case Index

Afghanistan, 116, 129

Bangladesh, 61, 142
Pultan Para, 26, 94

Bolivia, 116, 125, 139, 154

Brazil, 110, 129
Morada Nova, San Goncalo, Sume, 31, 120

Chile, 132
San Pedro de Atacama, 31, 48, 82, 89, 96, 101, 106, 141, 173

China, Meichuan, 25, 103

Ecuador, 143-144
Huerta Redonda, 173-174

Egypt, 43, 120, 167
Abu Raya, 25, 31, 52, 94

Ghana, Tono, 99

India, 22, 43, 51, 70, 87, 88, 90, 91, 99, 102, 106, 120, 123, 129, 132, 147, 148
Andhra Pradesh: MNC-TNC, 54, 69, 76, 84, 85, 92, 119, 168
Bihar, 47-48, 61, 69, 85
Tamil Nadu: Sananeri, 54, 63-64, 70, 76, 85, 90, 119, 135, 168-169

Indonesia, 27, 44, 49, 56, 73, 88, 115, 117, 121, 123, 126, 132, 137-138, 143, 144, 154, 157
Bima-Tabuyan, 31, 99
Pekalen Sampaen, 64-65
Subaks, 44, 72, 77, 88, 90, 94, 99, 102, 105-106, 115, 170

Iran, Nayband, 32, 47, 82, 89

Iraq, Daghara, 12, 48-49, 52, 57, 65-66, 82, 87, 172

Japan, 91

Kenya, 46, 116, 123, 156
Marakwet, 25, 32, 46, 51, 52, 61-62, 82, 103, 105, 173
Mwea, 7, 27, 31, 47, 67-68, 70, 76, 89,

91, 94, 99, 101, 107, 173

Laos, 62
Nam Tan, 49, 73, 82, 87, 171-172

Malaysia, 62
Muda, 24, 27, 65, 70, 72, 82, 94, 119

Mali, Office du Niger, 12, 123

Mexico, 13, 44, 54, 87, 120-121, 123, 126, 155
Diaz Ordaz, 48, 52, 73, 78, 85, 86
Oaxaca, 11, 50, 52, 69, 88, 115, 174

Morocco, 32, 110, 116, 129

Nepal, 13, 25, 44, 54-55, 84, 96, 98, 102, 104, 121, 134, 140
Argali, Chherung 12, 48, 49, 55-56, 70, 105, 169-170
Chattis Mauja, 8, 13, 46, 48, 51, 64, 67, 68, 70, 73, 76-77, 94, 106, 188, 169

Niger, 62, 73, 110, 129

Nigeria, 25

Oman, Izki, 32, 47, 49, 52, 86, 90-91, 172

Pakistan, 24-25, 32, 43, 44-45, 51, 55, 57, 65, 67, 9, 77, 99, 102, 123, 124, 135, 142, 143, 157
Chaj Doab, 31, 51, 85, 100, 103, 169
Daudzai, 51, 73, 88, 115, 169

Papua-New Guinea, Wamira, 51-52, 63, 103

Peru, 44, 83, 88, 91, 110, 115, 116, 124-125, 132
Quinua, 45, 48, 52, 55, 174

Philippines, 13, 23, 27, 30, 32-33, 44, 62, 86-87, 110, 111, 112, 116, 117, 120, 121, 123, 132, 134, 136, 138, 139, 141, 147, 148, 150, 151, 152-153, 155, 157
Buhi-Lalo, 24, 28-29, 65, 137, 170
Zanjeras, 8, 48, 62, 69, 72, 74, 96, 103, 125, 141, 170

Senegal, 50, 61, 77, 94, 110, 123, 155
Bakel, 73, 74, 94, 119
Matam, 70, 73, 94, 172-173

South Korea, 101, 110, 112, 123, 129, 140
FLIA, 70, 172

Southern Rhodesia. See Zimbabwe

Spain, 123, 133

Sri Lanka, 22, 27, 44, 52-53, 57, 62, 82-83, 100-101, 102, 104-105, 106, 119,

120, 134, 142, 144, 148, 150, 151, 152-153, 157
Gal Oya, 28-29, 31, 33-34, 57, 70, 71, 76, 84, 87, 90, 94, 112, 115, 119, 125, 126, 129- 130, 136, 139, 144, 147, 150, 156-157, 169
Minipe, 24, 34, 57, 74, 87, 112, 144, 150, 151, 156
Pul Eliya, 52, 56-57, 100, 169

Sudan, 43, 123
Gezira, 23, 25, 27, 31, 71, 91, 99, 173
Rahad, 23, 24, 31, 71, 88, 89, 81

Taiwan, 45, 62, 65, 70, 73, 77, 101, 102, 112, 119, 123, 155

Tanzania, Sonjo, 47, 52, 61, 82, 86, 87, 90, 101

Thailand, 44, 110, 123, 124, 129, 134, 141, 154
Nong Wai, 24, 131, 135, 141, 144, 150, 151
Seraphi-Sankaemphaeng, 48, 50, 51, 56, 171

Zimbabwe, Sabi Valley, 24, 99

Subject Index

Accountability of leadership, 87, 133-134
Acquisition of water, 38-39, 42-44, 54, 68, 84-85, 165
  effect on participation, 84
Activities in irrigation management, 37-57
Adequacy of water supply, 161, 163
Agency management, 7-8, 12, 13, 19-25, 50, 61, 67, 76, 86-87, 95, 105, 110, 146, 162
Agrarian Research and Training Institute (ARTI) (Sri Lanka), 139, 156
Agricultural extension. See Extension, agricultural
Agronomic practices, 99-100
Allocation of water, 38-39, 43-44, 54, 68, 110, 165
Application of water, 54
Asian Development Bank, 144, 150
Asian Institute of Management, 139, 153
Assembly mode of decision-making, 133, 134-135
Ateneo de Manila (Philippines), 139, 156
Ateneo de Naga (Philippines) 139

Bangladesh Agricultural Development Corporation (BADC) 26
Benefits, of irrigation, 3-4
  of participation, 24-26
  compared with costs, 27-30
Blueprint approach, 112, 130, 135, 154
Bribery, 54, 57, 85
Bureaucratic reorientation (BRO) 120-121, 145-159

Cadre, 113, 138
CARE, 125, 139
Cash, mobilization of, 48, 49, 50, 55, 98, 102, 110
Caste factors in irrigation management, 90, 102, 106
Catalysts, 134, 142, 144, 149- 151, 156
Chak, 24, 72
Collective action, 84-85, 93, 96, 146
Colorado State University, vii

Command Area Development Authority (CADA) (India), 148
Commercialization, 102
Committee mode of decision-making, 133
Communal Irrigation Committee (CIC) (Philippines) 111, 123, 153
Communication, 10, 40, 50-51, 53, 57, 98, 165-167
Conflict, 51-52, 77, 85, 140
  reduction of, 20, 22, 29, 33, 44
Conflict of interest, between farmers and agency, 23-24, 31, 95, 121-122
Conflict management. See Conflict resolution
Conflict resolution, 10, 11, 40, 51-53, 165-167
  women's role in, 92
Construction, 38, 41, 46, 167
"Cookie cutter" approach. See Blueprint approach
Coordination, 10, 40, 166-167
Cornell University (USA), vi-viii, 13, 139, 156
Corruption, 26, 54, 91, 106, 120, 152-153, 156
Costs of participation, 20, 26-27, 51
  comparison with benefits, 27-30
Criteria for evaluating performance, 31-32, 181-183
Cropping intensity, 20, 56, 141
Cropping patterns, 97, 99
Cultural factors in irrigation management, 77, 93, 102-103. See also Islam

Decision-making, 10, 40, 41, 46-48, 53, 165-167
  structure of, 133, 142
Demand for water, influence on participation, 99-102
Dependence on irrigation, 100-101
Design of water user associations, 127-144
Design of structures, 25, 38, 41, 45-46, 165, 167
  inadequate consultation of users in 120-121, 126
Dharma Tirta (Indonesia) 49, 73, 88, 170-171
Direct participation, 47, 68
Distribution of water, 22, 30, 38-39, 43-45, 54, 68, 162
Ditch-tenders, 64, 76, 90, 169, 170, 173
Donor role in participatory irrigation management, 50, 116, 153-155
Drainage, 3, 38-39, 43, 45, 53, 54, 55, 165

Economic factors, 93, 97-102
Economies of scale, 114
Efficiency in distribution of water, 126
Elites, 19, 48, 57, 85
Environmental impacts of irrigation, 21, 23, 31, 55
Equity in distribution of water, 20, 26, 30, 162-163, 164
  in investments, 143

Ethnicity, 81, 103, 106
Evaluation, 166
Executive mode of decision-making, 142
Existing organizations, for water management, 69, 115-117, 124-125, 127
Extension, agricultural, 95, 129-130, 140, 151
Extra-organizational activities, 53

Food and Agriculture Organization (FAO), 154, 157
Federation, of water user associations, 62, 74, 76, 114, 130-131
Fiscal crisis, of LDC governments, 154
Flexibility of water supply, 162
Flow of water, characteristics of, 161, 163
Ford Foundation, vii, 153
Formal organization, 10-11, 47, 64, 68, 71, 84, 98, 115, 124, 143-144. See also Informal organization
Free-riding, 114
Functions of water user associations. See Activities in irrigation management, and Water user associations, multi-purpose vs. single-purpose

Gender. See Women's role in irrigation management
Groups, for irrigation, 62, 72- 74, 76, 77, 88-89. See also Turnout groups, and Water user associations

Headman, for irrigation, 48, 49, 52-53, 56-57, 82, 88, 169, 170, 171
Heraclites, 96
Historical factors in irrigation management, 93-96

Ideology, of regime, 112
Incentives for users to participate in irrigation management, 83-86
agency incentives, 146-148
Indian Institute of Management, Bangalore, vii
Informal organization, 10-11, 47, 61, 64, 77, 124. See also Formal organization
Information, from farmers, 15, 48, 94, 126
In-kind, resource mobilization, 49-50, 68
International Irrigation Management Institute (IIMI), vii, 140, 157
Investment in structures, 101, 133, 146
Islam, influences on irrigation, 47, 52, 66, 89, 103

Joint management, 7-8, 13, 24, 162, 163

Kenya Tea Development Authority (KTDA), 156
Knowledge building institutions, 139

Labor, availability of, 101
mobilization of, 10, 24, 48, 49, 51, 68, 80, 125, 140, 147
Land tenure, effects on participation, 101-102, 105-106, 132-133
Landholding patterns, influence of, 69, 71, 78, 80, 85, 101-102, 105, 132, 141
Landless farmers, 80, 132-133, 141
Leadership, 86-87, 133-135, 142. See also Accountability of leadership
bureaucratic leadership, 148, 150-151, 152
payment of, 134-135
responsibilities of, 135, 142
selection of, 86, 88, 91, 115
Learning process, 29-30, 112-113, 137-138
Legal creation of water associations, 115, 123, 135-136
Levels of operation, 43-44, 59-78
Levels of organization, 10-11, 59-78, 110, 131, 134
one-level organization, 60-61, 70
two-level organization, 61-63, 70, 72
three-level organization, 13, 63-64, 70, 73
Location, advantages of head- end, 26
effects on participation, 79-80, 85-86
Local government, role in water management, 87-89, 92, 115, 129, 132

Main system management, 5, 88, 125-126, 146
Maintenance, of organizations, 29, 68, 83
of structures, 15, 20, 38, 41, 46, 47, 85, 110, 137, 149, 165, 167
lower status than design and construction, 152-153
Meetings, 134
Membership, basis of, 69, 80, 96, 105, 129, 131-133
Messengers, 51, 169, 171
"Modern", 52, 82, 87, 88, 124. See also "Traditional"

National Irrigation Administration (NIA) (Philippines), 22-23, 86, 87, 110, 111, 117, 123, 136, 138, 147, 148, 150, 152, 155, 157
Network, support group, 111-112
Nomenclature, for levels, 60, 67
Non-agricultural activities, implications for irrigation management, 81, 96, 105, 132

Objectives, of irrigation management, 19-24. See also Conflict of interest

Ombudsman role, 151
Organization, of farmers, 9-10, 40-42, 46-53. See also Levels of organization
  coordination with technical activities, 139
  focus of activity, 40-43, 46-53, 55, 168-174
  organizations as distinguished from institutions, 9, 13
Operation, 15, 38, 41, 47, 68, 110, 137, 149, 167. See also Levels of operation
  lower status than design and construction, 152-153
Outlet, 19, 43, 59, 72, 84-85, 110. See also Turnout
Overseas Development Institute, 161

Pareto optimality, 90
Parsons, Talcott, 166-167
Participation, farmer, 5, 8-9, 153-155, and passim
  changed relevance at different levels, 75
  in relation to water supply, 84, 91
  measurement of, 144
  optimization, not maximization, 26, 84, 91, 110
  strategy for introducing, 137-139
Part-time farming, 80
Patrols, 54-55, 56, 85, 168, 170, 172
Pilot projects, 137-138
Planning, 40, 166-167
Policy considerations, 4, 109-126
Political support, 103, 111-112
Politics, influence on water management, 82, 83, 93, 102-103, 118-119, 125
Population pressure, 15, 23, 30
Predictability of water supply, 20, 162, 163
Price, of crops, 100
  of water, 100
Primary level of irrigation, 60, 114
Private contractors, 139, 144
Private voluntary organizations (PVOs), 139, 144, 150
Production, objectives, 20-22, 120-121
Profitability, 100
Property rights, effects on incentives, 95, 117-118, 122, 136
Pumps, 12, 50, 61, 73, 78, 94, 95, 98, 104, 172-173

Qanats, 32, 47, 82, 98

Rainfed cultivation, 94, 100-101, 122
Rehabilitation, 8, 15, 27, 29, 33, 94, 128, 144, 147
Reliability of water supply, 20, 161, 163
Reservoirs, 25, 43, 103, 131
Resource management, 10, 40, 48, 165-167
Resource mobilization, 10, 20-21, 22-23, 28, 40, 48-50, 53, 56, 64, 71, 91-92, 165-167. See also Cash, In-kind, and Labor

"Retailing" of water. See "Wholesaling" of water
Richer farmers, 80, 85, 90
Roles in irrigation management, 79, 81-83, 86-87, 90, 168-174
Rotation areas, 65
Rotation of leadership responsibilities, 73, 86, 91

Scale, large and small, 7-8, 26, 27, 47, 59, 70-71, 75-76, 98, 123
Scarcity of water, 52, 97, 99. See also Supply of water
Seasons, variation, 33, 44, 98-99, 134
Shares, water rights, 44, 49, 50, 52, 55, 172
Size, of command area, 7, 62, 64-66, 69-71, 130-131, 141
Socio-technical process of irrigation, 3-6, 8, 60
"Software", 5, 28, 33
Soils, 3, 15, 97, 98. See also Drainage
Solidarity, 31, 76, 106
Source of water, 98
Specialization in irrigation management tasks, 32, 77, 81-83, 88, 90, 102, 115, 168-174
  in water user association formation, 149-151
SSRC (Social Science Research Council), vii, 140
Stealing of water, 43
Structures, focus of activities, 3, 9, 38, 41, 45-46, 54, 84
  social vis-a-vis physical, 9, 38-39, 104
Sufficiency of water, 97
Supply of water, effect on incentives, 83-84, 93, 97-99, 161

Tanks, 64. See also Reservoirs
Technical cooperation between water users and agency, 148-149
Technical knowledge of water users, 8, 13, 25, 32, 45-46, 120-121, 148
  not always adequate, 26, 82, 94
Technological solutions, 104-105, 127
Tenants, 25, 76, 80, 143
Tertiary level, 60, 67, 147
Timeliness of water supply, 161, 163
"Traditional", 47, 52-53, 82, 83, 87, 88, 92, 115, 116
Tubewells, 23, 55, 61, 142
Turnout, 7, 24, 54, 64. See also Outlet
Turnout group, 62, 72. See also Water user associations

User management, 7-8, 12, 13, 19, 44, 61, 123, 147, 162, and passim
  acceptance of, 120-121
Upstream advantage, 26, 44, 69, 79-80, 85, 90, 106
Unit command area (UCA), 60, 62, 70, 74, 114
USAID, vii, 50, 91, 124, 127, 153, 154, 155
Utah State University, vii, 60

Variability of water supply, 161, 163

Warabandi, 45, 55
Water, focus of management activities, 9, 38, 41, 42-45
Water producer associations, 142
Water user associations (WUAs) 28, 44, 49, 53, 87, 106, 118, 119, 124
  design of, 127-144
  duties of, 135, 142-143
  establishing, 27-29, 149-151
  multi-purpose vs. single purpose, 116, 128-130. See also Existing organizations, and Legal creation of WUAs
"Wholesaling" of water, 120, 125-126
Women's role in irrigation management, 80, 89, 92, 125, 132, 141
World Bank, 32, 77, 82, 110, 116, 124, 127, 129, 135, 155

Yields, crop, 20, 23, 29, 56, 95